OBJECT ORIENTED TECHNIQUES
IN
TELECOMMUNICATIONS

BT Telecommunications Series

The BT Telecommunications Series covers the broad spectrum of telecommunications technology. Volumes are the result of research and development carried out, or funded by, BT, and represent the latest advances in the field.

The series include volumes on underlying technologies as well as telecommunications. These books will be essential reading for those in research and development in telecommunications, in electronics and in computer science.

OBJECT ORIENTED TECHNIQUES
IN
TELECOMMUNICATIONS

Edited by

E.L. Cusack
Information Systems
BT Laboratories
Martlesham Heath
UK

and

E.S. Cordingley
Design and Performance
BT Laboratories
Martlesham Heath
UK

CHAPMAN & HALL

London · Glasgow · Weinheim · New York · Tokyo · Melbourne · Madras

Published by Chapman & Hall, 2–6 Boundary Row, London SE1 8HN, UK

Chapman & Hall, 2--6 Boundary Row, London SE1 8HN, UK

Blackie Academic & Professional, Wester Cleddens Road, Bishopbriggs, Glasgow G64 2NZ, UK

Chapman & Hall GmbH, Pappelallee 3, 69469 Weinheim, Germany

Chapman & Hall USA, 115 Fifth Avenue, New York, NY 10003, USA

Chapman & Hall Japan, ITP-Japan, Kyowa Building, 3F, 2-2-1 Hirakawacho, Chiyoda-ku, Tokyo 102, Japan

Chapman & Hall Australia, 102 Dodds Street, South Melbourne, Victoria 3205, Australia

Chapman & Hall India, R. Seshadri, 32 Second Main Road, CIT East, Madras 600 035, India

First edition 1995

© 1995 British Telecommunications plc

Printed in Great Britain at the University Press, Cambridge

ISBN 0 412 61460 X

A catalogue record for this book is available from the British Library

∞ Printed on acid-free text paper, manufactured in accordance with ANSI/NISO Z39.48-1992 (Permanence of Paper)

Contents

Contributors

S Cairns — Advanced Services Management, BT Laboratories

P G Coley — Software Engineering Centre, London

E S Cordingley — Design and Performance, BT Laboratories

E L Cusack — Information Systems, BT Laboratories

H Dai — Formerly University of Ulster at Jordanstown

A C T Drakeford — Software Engineering, BT Laboratories

R P Everett — Formerly Network Management, BT Laboratories

J A Graham — Software Development, BT Laboratories

A G C Heritage — Software Engineering Centre, London

P H J Houseago — Advanced Configuration Network Management, BT Laboratories

A J Judge — Formerly Network Management, BT Laboratories

P V Muschamp — Systems Engineering, BT Laboratories

B M Osborn — Formerly Network Planning Systems, Software Engineering Centre, London

J R Parker — Advanced Configuration Network Management, BT Laboratories

S Rudkin — Distributed Systems, BT Laboratories

C J Selley — Software Engineering Centre, London

R Shomaly — Formerly Service Management, BT Laboratories

C D Turner — Formerly Transaction Processing Systems, BT Laboratories

C D Wezeman — Advanced Specification, BT Laboratories

C T Whitney — Formerly Software Engineering Centre, London

J C R Wilson — Advanced Specification, BT Laboratories

P M Yelland — Formerly Intelligent Systems, BT Laboratories

Preface

'Object oriented' was one of the most controversial computing buzz words of the late 1980s. Some loved it, some loathed it, but the majority were simply confused. The 'paradigm shift' to object oriented programming languages was championed by proponents as the 'silver bullet' solution to all the ills of the software industry; it was mocked by others of equal eminence as an overhyped glossy repackaging of old ideas. A third much larger group of people were sidelined, mystified by the jargon and unsettled by the fact that experts were failing to agree amongst themselves on some of the most fundamental definitions. The mid 1990s finds principles of object orientation clarified and conceptual differences largely reconciled.

The technology is rapidly maturing. Tools are being developed to support object oriented analysis, design and development and the technology is being applied to solve industrial problems. Standards are being agreed. In the mid-1980s international standards bodies elected to experiment with an object oriented approach to structuring the information required to manage open systems interconnection networks. The object oriented concepts developed in this field are now allowing engineers to model live network problems.

The motivation behind the use of object technology is many threaded. Benefits claimed for object orientation (for example, see Coad and Yourdon [1]) include:

- increased understanding of the problem domain;

- improved ability to manage complexity;

- improved interaction among domain experts, analysts, designers and programmers;

- greater consistency across system development activities;

- greater exploitation of commonalities;

- systems being more resilient in a changing world;

- reuse both within the life cycle and in later developments.

It is the use of objects as the central focus of systems which brings these benefits. Object orientation lets computers model the world in a way that mimics human perception [1, 2]. This makes designing and reasoning about systems easier. Basing objects on real-world counterparts means that system components are recognizable to all stakeholders in a system. Objects provide a consistent set of concepts and representations that can be used throughout the life cycle, so there is less need for translation between teams working on different phases of system development. More effective scrutiny is thus possible; misunderstandings and misconceptions are reduced.

There is less to change through the life of a system as objects are relatively stable. Objects encapsulate together both the data and operations on that data, in such a way that the effects of errors and of change are localized. This eases testing, debugging, maintenance and enhancement of systems. It is the specialization and generalization of classes of objects that encourages reuse and greater exploitation of commonalities. Generalization and specialization gives object oriented systems an open/closed nature which helps future-proof systems and is particularly powerful in the control of change.

Once an object oriented system with its particular collection of objects is implemented satisfactorily, it can be regarded as closed to further radical change, and be deployed. It remains open, however, throughout its operational life to many kinds of change which can be made without impacting on the closed part of the system. The implementation of existing objects may be altered so long as they continue to behave as they are expected to, i.e. as defined by their interfaces. Additional classes of objects may be added. These may be specializations of existing classes or entirely new ones. This property of being simultaneously open and closed provides a flexibility unavailable in traditional systems.

As a huge producer of software, BT is naturally very interested in innovations and emerging best practice in software development. We have evidence within the company of the value of object orientation in managing the increasingly complex and dynamic telecommunications systems. Our experience to date has been positive and there is now every reason to be optimistic about the prospect of achieving higher quality systems as object technology enters the operational and commercial world. BT is a member of standards bodies and of major international groupings of companies — such as the Network Management Forum, the Object Management Group and the Telecommunications Information Networking Architecture Consortium — developing standards and producing innovative applicable solutions to industrial weight problems including the use of distributed systems.

This book publicizes a sample of BT's leading-edge work in object technology from research, standards and the application of object oriented solutions, to practical problems of telecommunications engineering. Several chapters derive from BT projects tackling issues in network and services management — from the use of AI technology in the laboratory, the design of an operational network routeing management system, to field trials of a real-time fault-management system. Others derive from projects into advances in software engineering clarifying concepts fundamental to the use of object technology in software, touching on topics such as systems analysis, formal methods, testing and project management. Two chapters explore the introduction of object oriented ideas into legacy systems implemented in COBOL.

The book has been divided into two parts which might loosely be regarded as research and development. The balance between the two, weighted towards the research activity, reflects the maturity of the technology.

In Part I, Chapters 1-8 contain accounts of a variety of research activities — analytical work extending current knowledge, illustrative case studies, investigatory applications, prototyping of systems using new ideas and development of demonstration systems.

To provide a common understanding of the basics, Chapter 1 provides a brief and, hopefully, accessible tutorial on object orientation which establishes a foundation for understanding the approach to object technology adopted and elaborated in the rest of the book. Chapter 2-10 discuss in some detail further object technology issues, illustrating discussions with examples from within BT.

A detailed explanation is provided in Chapter 2 of the open distributed processing (ODP) concepts of templates, types and classes, a set of concepts which has been notorious for generating confusion. By drawing from existing models, the ODP initiative has endeavoured to produce a consistent object model that includes the power of different interpretations. The chapter explains the background to the ODP definitions and explains why they provide a powerful general-purpose object oriented modelling framework.

Chapter 3 introduces the notion of roles and role restrictions in describing how Smalltalk was augmented with classification facilities like those found in knowledge representation language KL-One and its successors. The chapter discusses these facilities and describes how the 'term classification' enriched language was used to prototype an information base containing both large amounts of relatively simple data and small amounts of highly complex information.

Chapter 4 describes a model-based management approach to the experimental development of network service and customer administration systems within the ADVANCE European collaborative projects. Model-based management embodies the common information model which is a single logical entity containing the shared management information and functinality of the system. OBSIL, a high-level object manipulations language, used to develop the experimental system, is also described.

Chapter 5 is concerned with how theoretical ideas about object orientation can be adapted to address practical problems in commercial data processing. Its focus is the large number of 'legacy' systems implemented in COBOL and using network and relational databases. A method of re-engineering is developed which changes the structure of the existing code from a traditional to an object oriented design. For forward engineering, a four-model approach to analysis and design is described and recommended.

In Chapter 6 a deeper look is taken at the meaning and nature of encapsulation and a framework is developed for analysing the encapsulation strategy of systems. This approach is illustrated by applying it to one of BT's major systems under development, a transitional system influenced by both traditional and object oriented technology. Though new, the system was constrained to be implemented in COBOL and had many features of 'legacy' systems discussed in Chapter 5. Its design had, however, introduced a measure of object technology through encapsulation making its approach to encapsulation particularly interesting.

The need for more precise ways of specifying behavioural requirements is addressed in Chapter 7. The complementarity of object orientation and formal methods of object oriented analysis is argued, providing an appealing way of identifying the objects in the system, the use of the formal methods specification language Z producing a stronger, more precise abstract model of the intended behaviour of the system than the object oriented analysis on its own can achieve.

Presented in Chapter 8 are the results of a case study which investigated the use of object oriented Z, ZEST, for specifying managed objects and the subsequent implementation in C++ of these formally specified managed objects. Firstly, there is a description of current practice in managed object specification using the Guidelines for the Definition of Managed Objects (GDMO), an international standard modelling technique. Then, stereotyped mappings between ZEST and C++ structures are introduced and the effort required to implement from ZEST is compared with that required to implement from GDMO.

Chapter 9 describes the development of a working demonstration model to prove the viability of, and derive standards for, BT's Co-operative Network Architecture for Management with its two principle layers, the service management layer and the network management layer. The demonstrator emulates the virtual private network scenario, showing the provision and removal of network services. The software evaluates configuration interworking between service and network level management systems.

Chapter 10 examines the need for testing class-based, object oriented systems in a systematic way at the many levels of the 'V' model of the testing life cycle. It shows where the traditional approach to testing can and cannot be applied to object oriented systems. The special testing needs of object oriented systems are considered, together with what is required of the environments used to test them.

In Part II Chapters 11-14 tell of experiences using object technology on real developments within BT.

In Chapter 11 the impact of object orientation on the life cycle and management of the software development process is considered, drawing evidence from two large object oriented projects. It is argued that effective use of an object oriented approach requires a reassessment of the processes at all stages of the life cycle and a new approach to life cycle management is advocated. Experience also suggests there will be changes within the stage — conventional testing stages, for example, are likely to be merged into a stage called 'object integration testing'.

Chapter 12 describes the background to, the analysis and the specification of, a managed object model which defines a standard, generic interface for the configuration of services on digital switching elements on a telecommunications network. The adoption of such a standard will allow the telecommunications companies to provide new services with minimal development to their operational support systems. Companies gain vendor-independence, allowing network element procurement to be carried out on the basis of business and technical considerations such as quality of service, cost and functionality provided.

Described in Chapter 13 is the application of object oriented principles to the analysis, design and implementation of a major BT network management support system, the network routeing management system. The design of its digit decoder is used to illustrate the traditional and the generic object oriented approaches to design, highlighting differences and benefits resulting from the use of an object oriented approach and noting areas of unsatisfied expectation. The principal 'lessons' gained by the development team over the two-year period are presented.

The application of object oriented techniques to real time fault management is described in Chapter 14. In particular it addresses the identification of the single most probable cause of a group of alarms — the correlation of transmission network alarms. A model of the network was developed and stored using an object oriented database. Two approaches to the problem of correlation — 'out-of-model' and 'in-model' — are described. Reasons for choosing the latter for the local fault management system are given and early results from the system's field trial are reported.

The book's 'running order' takes the reader from the explanation of ideas, to their experimental use, then into real application. It also juxtaposes chapters with similar themes. Chapters 1-4 introduce concepts and models used in object technology. Chapters 5 and 6 focus on the constraints limiting the use of object technology because of existing 'legacy' systems. Chapters 7 and 8 address the use of formal methods within an object oriented approach. Chapter 9 uses managed objects, introduced in earlier chapters, in a demonstration system. Chapters 10 and 11 consider technical and managerial aspects of project management (respectively), both commenting on the testing of object oriented systems. Chapters 12-14, like Chapter 11, give practical experience of using object technology in BT, proceeding respectively from designing a generic interface, to design of a major system, to a system which has begun its feasibility field trial.

Starting at the beginning and going through to the end, however, is not recommended for all readers. Different routes through the book will be better for different readers. Everyone should start with Chapter 1, if only to appreciate the common base upon which other chapters build. Then recommended routes diverge.

Newcomers to object technology and to telecommunications (see Fig. 1) are recommended to proceed to:

- Chapter 2 then 4 if they are interested in open distributed processing;

- Chapters 3 and 4 followed by 6 (which is challenging) if they are interested in models behind object technology;

- Chapter 5 followed by 6 (which is challenging) if they are interested in re-engineering 'legacy' systems;

- Chapters 7 and 8, then 13 and 14 if they are interested in formal methods and managed objects;

- Chapters 10 and 11, followed by 13 and 14, if they are interested in project management;

- Chapters 4 and 12 for an introduction to telecommunications concepts, followed by 13 and 14 to develop an understanding of telecommunications applications;

- Chapter 9 should be saved until the reader has a good grasp of both object technology and telecommunications.

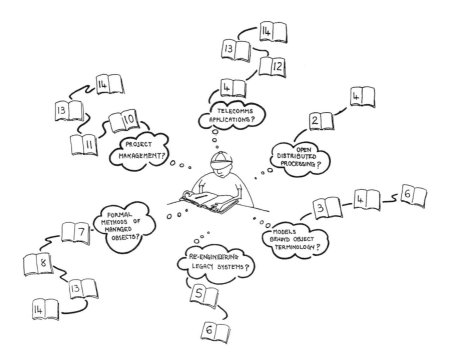

Fig. 1 Route for newcomers to OO and telecommunications.

Experienced academics and research software engineers (see Fig. 2) will be able to handle the challenges of Chapter 6 right away and that should be their next step after Chapter 1 if they are interested in encapsulation; Chapter 3 should be second for those interested in agents; Chapter 4 second for those interested in model-based object oriented design. Chapters 5, 7, 10 and 11 are particularly recommended to project managers. Chapters 4 and 8 (for background) and then 12-14 are recommended for those whose interests are telecommunications-specific.

Telecommunications technologists (see Fig. 3) may want to visit Chapter 8, which lays a foundation for understanding managed objects and their use, before going on to Chapters 4, 12-14, then Chapter 9.

Chapter 9 assumes knowledge of object oriented terms and concepts from Shlaer/Mellor [3], Rumbaugh et al [4] and Booch [5] which are not elaborated in the chapter nor included in the Chapter 1 brief overview. The reader may want to refer to those publications for additional explanation of concepts. Chapter 14 assumes some knowledge of telecommunications terms.

Fig. 2 Route for the experienced academic or research software engineer.

Fig. 3 Route for telecommunications technologists.

Many people contributed to the production of this book and we would like to thank them all. In particular, however, we are indebted to our external reviewers Ralph Hodgson formerly of IDE Ltd, now providing object oriented consultancy at IBM, and to Gordon Blair and his colleagues in the Computing Department at Lancaster University; they helped us maintain high standards of quality, readability and presentation.

E L Cusack
E S Cordingley

1. Coad P and Yourdon E: 'Object oriented analysis', (second edition) Yourdon Press/Prentice Hall International (1991).

2. Wegner P: 'Concepts and paradigms of object oriented programming', OOPS Messenger, $\underline{1}$, No 1 ACM Press (August 1990).

3. Shalaer S and Mellor J: 'Object lifecycles', Yourdon Press (1988).

4. Rumbaugh J et al: 'Object oriented modelling and design', Prentice-Hall International (1992).

5. Booch G: 'Object oriented design with applications', Benjamin Cummings (1991).

Part One

Research

1

OBJECT ORIENTATION IN TELECOMMUNICATIONS ENGINEERING

E L Cusack and E S Cordingley

1.1   INTRODUCTION

There is no single universally correct object oriented method of systems analysis, specification, design or programming. Object orientation is a chameleon. It encompasses a common core of generic concepts whose precise definitions can be tuned or specialized as necessary to provide what users require. The major traditions contributing to its growth include programming, database technology, knowledge engineering and formal methods.

The purpose of this chapter is to introduce and discuss the main concepts of object orientation and its use in telecommunications to provide a common understanding upon which the chapters that follow can build. We do not pretend that this is a thorough tutorial. There are now many excellent textbooks available, written by acknowledged authorities and aimed at different audiences. The interested practitioner may wish to select a book dealing with systems analysis (for example, the books by Coad and Yourdon [1] or Rumbaugh et al [2]), modelling and design (for example, books by Booch [3] Coad and Yourdon [4], Wirfs-Brock et al [5], and Martin and Odell [6]), programming (in languages such as C++ [7], Smalltalk [8], Eiffel [9]) or database technology (for example Hughes [10] and Oxborrow [11]). Research workers may gravitate towards the proceedings of major annual conferences such as Object Oriented Programming, Systems,

Lanuages and Applications (OOPSLA) [12] or the European Conference on Object Oriented Programming (ECOOP) [13]. Wegner has published a thorough account of the concepts and paradigms of object orientation, including a historical perspective [14]. The book by Blair et al [15] provides a readable general-purpose introduction to the various strands of object orientation, with chapters on basic principles, languages, distributed systems and applications. Oxborrow includes an excellent introduction aimed at the database community [11].

Section 1.2 of this chapter presents the fundamental ideas of object orientation, introducing amongst others the concepts of state, behaviour, encapsulation, type, class, inheritance, subtyping, and polymorphism. In section 1.3, the application of object oriented technologies is discussed, particularly object oriented modelling and the use of standards in telecommunications. Section 1.4 contains our concluding remarks.

1.2   FUNDAMENTAL CONCEPTS OF OBJECT ORIENTED TECHNOLOGY

This section discusses the concepts used to develop object oriented **descriptions** or **models** of real world systems. At many points in this section the term 'description' is used to mean either a program or a specification or model expressed in a non-executable language. The reason for doing this is that the distinction between executable and non-executable descriptions is largely irrelevant to a discussion of fundamental concepts.

Wegner [16] has documented the philosophic evidence for the existence of a human cognitive filter based on abstraction and classification, techniques which are fundamental to the object oriented approach. This suggests that object orientation reflects a deep and natural human perception of the world; it therefore seems appropriate to begin the introduction to object orientation using a natural example.

1.2.1   Objects

An **object** is an abstraction or model of a discrete component or 'building block' of a system. The system is the universe of discourse, i.e. what is being talked about — suppose, for example, that the system is the family home; the components that might relate to the home include various items of furniture, electrical equipment, household finances, the family and household pets. Four of these which provide some variety might be a table, a kettle, the mortgage and a dog. The cognitive filter provides mental models of these real world objects in a way that lets us clearly distinguish between them.

1.2.1.1 Encapsulated state and behaviour

The mental models have some important features in common. First of all, they are discrete — the boundaries of each item can be clearly recognized. In some sense, therefore, each of the four objects is encapsulated.

It can be spoken of as a single whole thing. Some of what the object is like, its **state**, and the way it works, its **behaviour**, is apparent to the human observer, but some is not. Within the boundaries of objects, there are degrees of internal working which are observable, ranging from none (the table) to a complex physiology (the dog). But many of the internal workings of each object are not immediately apparent — the electrical process when the kettle is switched on, the work processes at the building society with whom the home is mortgaged, the dog's cardio-vascular system, whether or not it is hungry. These encapsulations show some but hide other information and processes. For a detailed consideration of encapsulation, see Chapter 6.

Each object, furthermore, **behaves**, or **interacts with its environment**, in a prescribed and understood way. This prescribed way of interacting is often called the object's **interface** with the environment. The position of the table in the room can be moved, water in the kettle can be boiled, and the dog fed. These interactions may alter the **state** of the object (the location of the table, whether the kettle is empty or full, how hungry the dog is) but they

..HOW HUNGRY THE DOG IS...

leave the **identity** of the object unchanged. In fact, the state of the object cannot be changed in any way other than by the possible behaviour of the object. (When we suspend this requirement, we enter the realms of the supernatural, where tables can fly through the air of their own volition and animals can speak!)

..TABLES CAN FLY THROUGH THE AIR OF THEIR
OWN VOLITION AND ANIMALS CAN SPEAK!

The packaging together of state (data) and behaviour (process) in an encapsulated description is a key characteristic of object orientation. Some objects (for example, the dog) initate their own interactions. These are often termed **active** objects in the literature. Others (for example, the table) depend on the environment to take the initiative. These are often termed **passive** objects.

1.2.1.2   Persistent identity

An object has a lifetime. The table and kettle were manufactured, and will be scrapped; the mortgage came into being on the day funds for the home were advanced, and will eventually be paid off. During its lifetime an object has a persistent identity that is unaffected by its interactions with the environment. Even if the dog is sold to a new owner who changes his name from Spot to Fido, it is still the same dog.

In summary, an object is an encapsulated component with a persistent identity; its state can be altered according to prescribed interactions with its environment, but in no other way.

IT IS STILL THE SAME DOG.

1.2.1.3   Message passing and polymorphism

A typical way of getting an object to do something is to 'send' it a 'message'. The sending of a message is usually stimulated by some event. That event might be an input from a user of the system, the passing of time, or a change in the state of some object.

Object oriented systems exhibit **polymorphism**. This means that sending the same message to different objects could stimulate different behaviours. Each object interprets the message in a way which is appropriate for itself. Closing an account and closing a door, for example, might both be regarded as sending the message 'close' to an object (the door or the account). The processes which resulted from the objects receiving such a message would clearly be very different.

1.2.1.4   Language and abstraction

When a system is described, irrelevant detail is suppressed. This produces a description which is an **abstraction**. The dog's owner knows that Spot is a two-year old Dalmatian; the Kennel Club judge will explain why it is not a perfect example of the breed (citing features that may not be detected by the untrained eye); if Spot is sick, the vet will reach a diagnosis, drawing on his or her knowledge of canine physiology. So an object description needs to be an abstraction which reflects the likely needs of users of the system of which it is part. The precision of the description too must be appropriate. Use of the term 'description' in this chapter depends on an intuitive notion of the terms in which a description might be couched. In fact, a description must be expressed in some **language** — loosely, a collection of concepts and a notation in which they can be expressed, together with rules for using the notation (syntax) and rules for deriving the meaning of descriptions (semantics).

The concepts of a language determine the sort of thing that the users of the language can describe well, or at all. For example, it would be much harder to reason about electronic circuits without the usual engineering notational conventions of circuit diagrams. A language, the use and interpretation of which is governed by rules which are rigorously — perhaps even mathematically — defined, is said to be **formal**.

1.2.1.5   Describing objects

An object should be described in an appropriate way, using a language fit for the intended application of the description. To be able to describe objects, therefore, a language must have concepts of state and behaviour.

State is most commonly defined in terms of **attributes** or **variables** which can take values of a declared kind in a declared range. For example, the state of a kettle can be defined by three variables, one indicating whether or not the power supply is switched on, another indicating the volume of water (say, between 0.0 and 1.0 litres) and a third indicating whether the water it contains is at boiling point or not.

Behaviour is often defined in terms of metaphors such as **message passing** between the object and its environment (see above), or equivalently the **operations** that can be performed on the object. Objects are described at this level of abstraction in terms of their interface with the environment — we specify the operations and their effects. For example, it could be explained to a young child that the operation 'switch on power', when applied to the kettle, results in the water being heated.

Object oriented programs are formal descriptions, but contain far more detail than a formal specification would. Objects are described at this level of abstraction in terms of how the interface with the environment is implemented. For example, an older child or adult understands that the operation 'switch on power' causes electric current to flow through the element of the kettle, and the resistance of the element results in the generation of heat. In some object oriented programming languages, objects are said to possess **methods** (chunks of executable code) which implement the object's response to messages received. These methods support the object's interface with the environment but the interface definition does not depend on any particular choice of implementation. This means that the code which implements the object's methods may change without the environment being aware of the change, provided the object behaviour continues to be as specified in the interface description.

1.2.2   Classes and types

The previous section began with the remark that the objects related to a home might include various items of furniture, electrical equipment, household finances and household pets. Spot was also described as a Dalmatian. This bears out a point made at the start of the section — it is impossible for humans to discuss their environment without abstracting and classifying. We know what a table is (or a kettle, or a mortgage, or a dog) and can describe one without reference to any individual object. Such an abstract description defines the common features of a collection of objects. This collection is called a **class**. We often describe individuals by first referring to their class ('Spot is a dog') and then adding extra distinguishing information ('Spot is a two-year old Dalmatian who likes to go for long walks'). We say that the object is an **instance** of a class to which it belongs. Creating a particular individual object within a system from a description of its class is called **instantiation**. Doing it while a system is operating is called **run time instantiation**.

SPOT IS A DOG.   SPOT IS A 2YR OLD DALMATION WHO LIKES TO GO FOR WALKS.

The **type** of an object is a partial description of its features. The object is said to be 'of the type' if it satisfies the type description. For example, many tables belong to the type made_of_wood. It is not hard to see that we can associate a type (the common features of the class) with each class. Conversely, each type determines a set of objects, though we may not require that set to be a class in our system description. The notions of class and type are intimately connected, and the reader might be forgiven for wondering if both are needed. Indeed, any conventional object oriented language or approach will most likely include either class or type as features, but not both. However, an understanding of both concepts is necessary in order to appreciate the subtleties of subtyping and inheritance.

Before moving on, we need to reconcile a difference in terminology between the conceptual modelling framework presented in this chapter and some well-known methodologies and programming languages, for example Coad and Yourdon [1] and Smalltalk [8]. In Coad and Yourdon 'class' is used to mean the definition of what the members of a collection have in common — what has been referred to here as 'type'. In Smalltalk the word 'class' is used to denote a distinguished object containing a 'template' description of a collection of objects and endowed with the ability to create new objects matching the 'template' on request. The 'template' is actually a module of code, and so the objects created by the class are all implemented in the same way. In such languages, the notion of class is therefore tightly bound up with the creation of new objects. The term 'class' as used in this chapter corresponds to the collection of all instances created by a Smalltalk class. The term 'type' as used in this chapter corresponds to an intuitive, implicit notion of the interface provided by objects created by the Smalltalk class. For casual discourse it is usually acceptable to use the terms class and type interchangeably [3] and this often occurs in the literature causing confusion when the distinctions need to be maintained. See Chapter 2 for a detailed discussion of these terms from an open distributed processing (ODP) perspective.

1.2.2.1   Subtyping

Some classes contain other classes (in the set-theoretic sense), denoting a 'kind of' relationship. In other words, objects may be instances of more than one class. For example, Spot is an instance of class Dalmatian as well as of class Dog. This allows amongst other things, objects to be described efficiently at the required level of detail. For example, a Dalmatian is a kind of dog, which in turn is both a kind of mammal and a kind of household pet. This 'kind of' relationship is called subtyping, and the (type associated with the) class Dog is a **subtype** of the (type associated with the) class Mammal, or alternatively that Mammal is a **supertype** of Dog. It is possible and natural for some classes to be subtypes of more than one other class. The subtyping relationship guarantees that an object of the subtype can be successfully substituted into an environment expecting an object of the supertype. It is sometimes useful and efficient to define a class with no immediate instances,

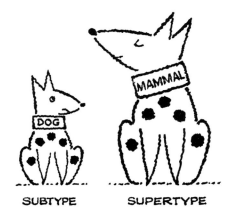

SUBTYPE   SUPERTYPE

i.e. each instance of the class is also an instance of a subtype of the class. Such classes are called **abstract** or **virtual** classes. For example, Furniture is an abstract class, since an object cannot be furniture without also being a table or a chair or a bed, or some other item.

Subtyping can be as sophisticated as the user wishes to make it, within the constraints of the language being used. The choice and definition of classes are important creative design decisions. If we are concentrating on static properties of objects, then it is relatively easy to define sensible subtype relationships. For example, Fig. 1.1 shows a simple **subtype hierarchy** (classification scheme — also referred to in the literature as a **class hierarchy**, a **generalization/specialization hierarchy** and an **aggregation hierarchy**) for furniture.

On the other hand, if we wish to describe objects in terms of their behaviour as well as their state, then it can be harder to tell whether or not one class is a subtype of another. The decision requires a well-defined and precise method of comparing the behaviour of one type with another. Some current object oriented programming languages support subtyping (for example, Eiffel [9]); others (for example, Smalltalk) do not. The problem has also been tackled by different researchers, using formal methods (for example Cusack [17]) or knowledge representation techniques (see Chapter 3).

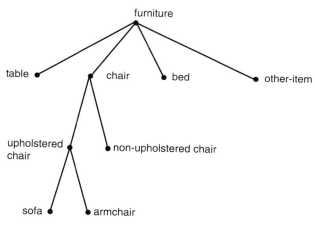

Fig. 1.1   A simple subtype hierarchy for furniture.

1.2.2.2   Inheritance

It is often efficient to describe a new type of object in terms of changes and additions to an existing type or types (**parent types**). For example, an electric kettle can be thought of as a jug made from heat-proof material, with the addition of a heating element and a 'switch on power' operation. This incremental approach, specializing a type to define its **child**, is known as **inheritance**. There is no single correct definition of inheritance — it is up to the language designer to decide what is likely to be useful and used. Inheritance is essentially a technique for producing system descriptions in a flexible yet structured way, reusing descriptions already produced.

It may be possible to cite only one parent type (**single inheritance**) or to cite several parent types (**multiple inheritance**). In the case of multiple inheritance, there may be a need to reconcile differences between the parent types — for example, the same name may be used by each parent type to denote a different operation. It may be compulsory to keep all existing interactions unchanged, or it may be possible to redefine or even delete some existing interactions. Thus a type derived by inheritance will be 'like' the parent type or types in some way [18], but need not — and, in many cases, will not — be a subtype. Similarly, a parent type may be, but need not be, a supertype. If inheritance is so tightly constrained that any class derived must be a subtype, then it is referred to as **strict inheritance**. For example, the programming language Eiffel [9] features strict inheritance. These definitions are far from academic — they have a powerful influence on the flexibility and expressiveness of the language.

1.2.2.3   Using subtyping and inheritance together

Inheritance and subtyping are still major research topics in object orientation — there is a detailed discussion of many of the issues in Wegner [14]. A paper by Lalonde and Pugh [19] illustrates the distinction between inheritance and subtyping with many examples. Porter [19] has explored the issue of separating subtyping in programming languages from implementation inheritance, concluding that both hierarchies can and should be supported.

In objected oriented programming languages such as Smalltalk, inheritance is an implementation technique. New classes can be defined by augmenting (inheriting from) classes already implemented. This is simply a special case of the more general definition of inheritance given above. Classification facilities, of the kind discussed in Chapter 3, are based on the notion of the concept, which closely resembles the notion of type defined in Chapter 3. By drawing on the properties of concepts, such facilities are able to determine automatically that one concept is a subtype of another, or that an object is an instance of a concept. The system described in this chapter combines these abilities with those of Smalltalk to produce a range of new capabilities not normally found in object oriented programming languages.

1.2.3   Part-whole relationship

A third important hierarchy used in object technology is the **part-whole** tree defined by the 'part of' relationship. It is a tree of objects and their component parts, referred to by Booch [3] as the assembly structure. Subtyping and inheritance hierarchies, by contrast, capture relationships between types.

The main advantage of a part-whole hierarchy is that it allows meaningful composition relationships to be captured involving objects whose types may bear little or no relation to each other. A kettle, for example, is a complex object with a number of components. It is made up of a body, a handle, a lid, and (if it is an electric kettle) a heating unit. These in turn may be regarded as being made up of smaller parts — the body is made up of a container and a spout; the lid may have two components, the body of the lid and a knob; the heating unit may be regarded as made up of a heating element, an insulating ring and a socket.

As with subtype hierarchies, the level of detail and the nature of the decomposition depends on how the structure is to be used. For example, for the purposes of drawing, it may be enough to think of Spot, the Dalmatian, as made up of four legs, a body and a head. For biological purposes, it may be more useful to regard Spot as having components such as organs, fat, fluids and so on, each of which can be viewed as having components down to the cell level. For micro-biological purposes, cells are themselves seen as complex units.

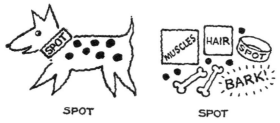

SPOT   SPOT

The scheme of parts need not be those which are needed to construct the whole in an engineering sense, i.e. parts which can be bolted, glued or otherwise put together. We are not at the stage where we could think of constructing Spot from his observable or even his biological parts. Furthermore, if Spot were to be flown to a new home he might be considered as made up of hunger, thirst, weight, and volume. Though these are legitimate components, useful in the air transport context, these are not parts for constructing a physical entity in the engineering sense.

The implication of an object being in a part-whole hierarchy is not standardized. Some systems are developed with a co-existence of parts and wholes. This means that the part and the whole are created in the system at the same time and are removed from the system at the same time — in other words, if the whole is removed, all of the parts are also removed. This is not necessarily the best strategy. An electrical repair firm may want to keep track of kettles and of kettle parts. It may want, for example, to salvage heating elements or lids from kettles whose spouts or bodies have been broken. In some systems the parts must be in the same place to make up the whole and in other systems parts can be far apart and still be part of the whole. The mortgage may be part of a building society system where repayments are made in one place, its interest rate determined somewhere else, neither of which is necessarily the place where the original of the signed document is kept. Chapter 6 discusses these notions of co-temporal existence and co-location in the context of encapsulation strategies.

1.2.4   An emerging international standard

Countless numbers of people have been confused by the many different definitions of object oriented concepts to be found in the literature. Booch, for example, is reluctant to distinguish between class and type [3]. In recognition of this problem, an international standards project on open distributed processing (ODP) has recently formulated a rigorous and consistent set of definitions related to class, type and instance [21, 22]. They are intended for application primarily to the modelling of distributed systems architecture but are not specific to that area. Chapter 2 explains the background to the ODP object oriented definitions and explains why they provide a powerful, general-purpose object oriented modelling framework. Section 1.3.1 below expands on the relationship between ODP modelling and other industrial initiatives.

1.3   APPLICATIONS OF OBJECT ORIENTED TECHNOLOGY

The applications of object oriented technology in communications engineering fall into two main groups:

● modelling and specification of communications entities, especially network and service elements;

- development of object oriented software.

The chapters contained in this book reflect both aspects. Of course, software development encompasses modelling and specification, but there are important areas where models and specifications are significant end-products in themselves (most notably in international standardization.)

1.3.1   Object oriented modelling

Models or specifications are high-level system descriptions which concentrate on what is required, not how the specification is implemented. They are an essential part of systems and software engineering. For example, such descriptions can be checked for completeness and internal consistency, and agreed with system end-users before being passed to implementers (see Chapters 7 and 8). Specifications can form pre-competitive international standards to open up markets, as exemplified by the international industry consortium, the Network Management Forum (NMF). They can help us better understand legacy systems already in place (see Chapters 5 and 6). They can also provide a route to rapid prototyping (see Chapter 3).

1.3.1.1   Managed objects

The objects most frequently discussed throughout this book are **managed objects**, abstractions of data communications resources, formulated for the purpose of open systems interconnection (OSI) network or services management [24]. A managed object is the management view of a resource, such as a modem, a circuit, a PC, a customer or a call. In other words, a managed object is an abstraction of an actual resource formulated for the particular purpose of management. There is an international standard modelling technique, the Guidelines for the Definition of Managed Objects (GDMO) [25, 26]. Chapter 8 gives more details, beginning with a description of current practice in managed object specification using the GDMO, and Chapters 9, 12 and 14 describe the use of the GDMO to tackle problems in network and services management architecture, services configuration on digital switches and transmission network fault diagnosis respectively.

The GDMO object oriented concepts specialize on the generic modelling framework for ODP mentioned earlier in section 2.4. They are also closely aligned with the object model created for object oriented software development by the international industry consortium, the Object Management Group (OMG), of which BT is an end-user member. A recent joint study by the Object Management Group and the Network Management Forum has begun to explore ways of reconciling the few differences with a view to speeding up the development of industrial strength network management products [27].

1.3.1.2 Formal approaches to specifying behaviour

A problem with the GDMO is that it provides only natural language English for the specification of managed object behaviour. Natural language descriptions of behaviour are often found in practice to contain unintentional ambiguity and other flaws which reduce their value. The benefits of enhancing the GDMO by formalizing managed object behavioural specification has been recognized by BT and other organizations, and moves in the international standards community have begun. Chapter 8 reports progress in using an object oriented variant of the formal specification language Z [28] for this purpose. Chapter 7 also explores behavioural specification in Z, showing how object oriented analysis techniques can be used in conjunction with Z. Chapter 4 describes how an analogous problem of behavioural specification was tackled by the European RACE programme in the context of advanced information processing (AIP).

Although a considerable improvement on natural language, Z is less expressive than the powerful process calculi (such as Milner's Calculus of Communicating Systems [29]) developed in the 1980s for modelling and analysing behaviour. The introducion of such behavioural specification ideas into object oriented modelling is still a research issue — in particular, it has proved difficult to reconcile process calculi with inheritance [30, 31].

1.3.1.3 *Post hoc* object oriented modelling

A third and younger strain of object modelling involves the application of object oriented constructs to non-object oriented systems such as BT's large existing database systems and object oriented use of non-object oriented languages such as traditional COBOL as discussed in Chapter 5 and the transitional system discussed in Chapter 6. These modelling activities can be used now, before the Object Oriented COBOL Task Group (OOCTG) of the American National Standards Institute (ANSI) arrives at standards for an object oriented COBOL (OOCOBOL) [32]. They will continue to be needed after such a standard is agreed because reprogramming or replacement of the major legacy systems, such as the customer services system, written in traditional COBOL, is prohibitively expensive.

The least intrusive approach involves identification of object-like patterns in existing code. These can be the units around which code can be reorganized, highlighting duplicated code which can be removed. They can be the basis for a new logical representation of the code, one which is more object oriented.

More intrusive, but acceptable in new systems or when parts of existing systems are being rewritten, are the techniques for using even traditional COBOL in a more object oriented way.

1.3.2   Object oriented software development

Object orientation not only affects the models and the applications themselves, but also their assessment — verification, validation and testing (see Chapter 10) — and the process by which they are developed (Chapter 11). Chapter 13 draws all the threads together in a description of the design and development of a new network routeing management system for BT's digital exchanges.

1.4   CONCLUSIONS

This chapter has introduced object orientation in communications engineering and should help to guide the reader through the remaining chapters. Many references have been given to books and papers which provide more detailed tutorial material.

Object oriented technologies allow modularizing and structuring of system descriptions in a consistent, intuitive way. Objects are a natural unit of complexity, so object orientation greatly improves our ability to describe complex systems. Object oriented approaches are flexible and encourage reuse. However, in some respects (particularly testing) object orientation remains immature.

It is helpful to review and summarize the main concepts of object orientation introduced in this chapter:

- an **object** is the encapsulated state and behaviour of a uniquely identifiable individual component of a system with a persistent identity;

- an object's **state** commonly defined in terms of **attributes** or **variables,** can be altered according to prescribed **interactions** with its environment, but in no other way;

- an object's **behaviour,** as observable by its environment, is prescribed by its **interface;**

- the way an object's behaviour (known as **methods**, or **operations**) is implemented is usually not known to its environment and can be changed without impacting upon its environment, provided the interface remains the same;

- an object's response to a stimulus (e.g. a **message**) is appropriate for that object — two different objects may respond differently to the same message, a feature of the technology which is called **polymorphism**;

- any object description is an abstraction, with irrelevant detail suppressed;

- an object should be described in an appropriate way, using a language fit for the intended application of the description;

- a **class** is a collection of objects with significant features in common. The description of these common features constitutes the **type** associated with the class; an object which is one of that collection is said to be an **instance** of that class;

- objects can be classified by their type using **subtype** relationships;

- **inheritance** is a technique for producing system descriptions in a flexible yet structured way, reusing descriptions already produced — in programming languages, therefore, inheritance is an implementation technique encouraging reuse of code already written;

- complex objects can be decomposed into component objects using part-whole relationships.

In communications engineering, object oriented modelling is being successfully applied to problems in OSI network management, network and services management architecture, services configuration on digital switches and transmisson network fault diagnosis. Object oriented software for operational use is now being designed and developed, and the problem of testing object oriented software is being addressed. Possibilities are emerging for re-engineering legacy systems written in COBOL using object oriented ideas. Promising longer-term directions include draft international standard definitions of object oriented concepts, improving the expressive power of current programming languages and the combination of object oriented analysis with formal specification techniques.

There is now every reason to be optimistic as object oriented techniques start to leave the research laboratory and enter the operational and commercial world.

REFERENCES

1. Coad P and Yourdon E: 'Object oriented analysis', (second edition) Yourdon Press/Prentice Hall International (1991).

2. Rumbaugh J, Blaha M, Premerlani W, Eddy F, and Lorensen W: 'Object oriented modelling and design', Prentice Hall (1991).

3. Booch G: 'Object oriented design with applications', Benjamin/Cummings Publishing Co (1991).

4. Coad P and Yourdon E: 'Object oriented design', Yourdon Press/Prentice Hall International (1991).

5. Wirfs-Brock R, Wilkerson B and Weiner L: 'Designing object-oriented software', Prentice Hall (1990).

6. Martin J and Odell J: 'Object oriented analysis and design', Prentice-Hall (1992).

7. Stroustrup B: 'The C++ Programming language', (second edition) Addison-Wesley (1991).

8. Goldberg A and Robson D: 'Smalltalk-80: the language and its implementation', Addison-Wesley (reprinted 1985) (1983).

9. Meyer B: 'Object oriented software construction', Prentice-Hall International Series in Computer Science (1988).

10. Hughes J: 'Object oriented database', Prentice Hall (1991).

11. Oxborrow E: 'Databases and database systems: Concepts and Issues', 2nd Edition, Chartwell Bratt (1989).

12. 'ACM SIGPLAN Notices', OOPSLA'92 Conference Proceedings, 27 , No 10 (October 1992).

13. Lehrmann Madsen O (Ed):'ECOOP'92 European Conference on Object Oriented Programming', Lecture notes in Computer Science 615, Springer (1992).

14. Wegner P: 'Concepts and paradigms of object oriented programming', OOPS Messenger, 1 , No 1, ACM Press (August 1990).

15. Blair G, Gallagher J, Hutchinson D and Shepherd D (Eds): 'Object oriented languages, systems and applications', Pitman (1991).

16. Wegner P: 'The Object Oriented Classification Paradigm', in Shriver B and Wegner P (Eds): 'Research Directions in Object Oriented Programming', MIT Press (1987).

17. Cusack E: 'Inheritance in object oriented Z', in America P (Ed), 'Proc European Conference on Object Oriented Programming', Lecture Notes in Computer Science 512, Springer (1991).

18. Wegner P and Zdonik S: 'Inheritance as an incremental modification technique, or what like is and isn't like', European Conference on Object Oriented Programming, Norway (August 1988).

19. Laldone W and Pugh J: 'Subclassing ≠ Subtyping ≠ Is-a', Journal of Object Oriented Programming, 3, No 5 (1990).

20. Porter H H III: 'Separating the subtype hierarchy from the inheritance of implementation', Journal of Object Oriented Programming, 4, No 9 (1992).

21. CCITT Rec X.902 — ISO/IEC CD 10746-2.1: 'Basic Reference Model of Open Distributed Processing — Part 2: Descriptive Model', (1992).

22. Taylor C J: 'Object Oriented Concepts for Distributed Systems', to appear in Computer Standards and Interfaces.

23. Willetts K and Adams E: 'Omnipoints — a way through the maze of open management', in Advanced Networking Week, Blenheim Online (1992).

24. Smith C C: 'OSI Systems Management: description and key role of CCITT (X.700 series) and ISO/IEC development', (1990).

25. Jeffree T: 'GDMO: Tools for defining OSI managed objects', in Advanced Networking Week, Blenheim Online (1992).

26. CCITT Rec X.720 - X.722 — ISO/IEC IS 10165 (Parts 1-4) 'Management Information Services — Structure of Management Information', (1992).

27. Ashford C (Ed): 'Comparison of the OMG and ISO/CCITT Object Models', report of the Joint Network Management Forum/OMG Taskforce on Object Modelling (February 1992).

28. Spivey J M: 'The Z Notation: A reference Manual', (second edition) Prentice-Hall (1992).

29. Milner R: 'Communication and Concurrency', Prentice Hall (1989).

30. Cusack E: 'Refinement, conformance and inheritance', Formal Aspects of Computing, 3, No 2 (April-June 1991).

31. Rudkin S: 'Inheritance in LOTOS', in Parker K R and Rose G A (Eds): 'Formal Description Techniques', North Holland (1992).

32. Topper A: 'OOT and COBOL: how do they fit together?', Object Magazine, 2, No 6 (March-April 1992).

2

TEMPLATES, TYPES AND CLASSES IN OPEN DISTRIBUTED PROCESSING

S Rudkin

2.1   INTRODUCTION

Over the next few years we will see the importance of both distributed and object oriented systems rise dramatically. Since a system that comprises a single machine is a special case of a distributed system, and since object orientation plays such an important part in distributed system design, it clearly makes sense[1] to adopt a common object model for all object oriented and distributed systems development. Open Distributed Processing (ODP) is developing that model. ODP is an international standards activity developing an object based framework for the design and construction of all kinds of multivendor distributed systems. ODP's object model lies at the heart of this framework.

Anyone who is familiar with a selection of object oriented programming languages will be aware that there are conflicting interpretations of many basic object concepts. The ODP descriptive model [1], which is already substantially complete, provides a single consistent set of concepts.

This chapter describes a small part of ODP's descriptive model — namely, the concepts of template, type, and class and any associated concepts. Section 2.2 provides some background on ODP, followed by a brief overview

---

[1] The main advantage of such a move will be improved skills portability.

in section 2.3 of the role of objects within ODP. Then, beginning with templates in section 2.4, the various concepts are presented in order of increasing complexity and dependency — for ease of reference, definitions are presented in Table 2.1 in section 2.4. The chapter concludes with a summary of the clarifications that have been made.

2.2 BACKGROUND ON ODP

Consider the information technology (IT) infrastructure of any large business today. More often than not, it will comprise many separate systems built from incompatible technologies. This situation has arisen because of the limitations of past technology, owing to the lack of interoperability between proprietary products, and because system procurers have focused only on local problems. This has led to the independent development of local solutions which are optimized to meet local requirements but have ignored company-wide considerations.

However, over the last decade, the emergence of distributed processing, based on *de jure* and industry standards, has overcome many of the past technological limitations. Moveover, processing, memory and communications are both cheaper and faster. Building on this push from advances in technology, there is now a strong pull from business. The business community has realized the importance of fast access to business critical information [2]. Being first in the global market means being able to reach the right information at the time it is needed. Increasingly companies are using fast access to information to improve their efficiency and to assist the move towards a flatter, less hierarchical corporate structure.

Businesses are also discovering the benefits of closer co-operation between supplier and consumer. Streamlined procedures governing their interaction are often most effective when supported by their IT systems. Interconnection of their IT systems can provide a simple and effective means for despatching orders, paying bills, checking the availability of services, etc. Whatever the business requirement for interconnection, an open and widely accepted architecture, governing interconnection, is required.

The central requirement that such an architecture must address is evolution. As businesses evolve, their IT requirements change — this is especially true for large distributed systems. Often such changes will involve integrating independent and technologically diverse systems, resulting in a heterogeneous system (i.e. a system which is based on a mixture of different hardware, operating systems and programming languages). A related requirement, then, is the need to accommodate heterogeneity, rather than imposing homogeneity. This implies that support for interworking and portability across a wide range of technologies must be provided.

A number of architectural principles are shown in Herbert [3] to meet these requirements. Foremost amongst these is an object based approach. In the object based world, systems are described as a collection of objects. Each object is an identifiable encapsulated entity that provides one or more services that can be requested by clients. Client objects can ask for services by sending a request to a chosen server object. Each object can be thought of as a separate logical process with its own state, which may be changed by interacting with other objects.

Objects are desirable because of their two key properties — abstraction and encapsulation. Abstraction is the process of ignoring inessential details for the job in hand. In a programming environment, objects are typically used to hide implementation detail. Encapsulation is the property of enforcing an abstraction. It hides the internals of an object so that clients interacting with the object cannot see the mechanism (algorithm) being used. Abstraction plus encapsulation helps to achieve interworking and portability. Together they deny the possibility of establishing implementation-dependent interactions. This encourages well-defined boundaries between system components, which in turn leads to the development of reusable services.

A number of industry initiatives are adopting an object oriented approach to meeting the requirements raised above. The most significant of these are the Object Management Group's CORBA [4] and ODP. ODP is providing an overall framework, called the ODP reference model [1, 5-7], for the development of heterogeneous distributed systems. A major influence on the ODP architecture has been the object-based architecture developed by ANSA [8]. Now that the reference model is maturing, ODP has started to initiate work on the specification of individual components. The OMG CORBA can be likened to a core collection of these ODP components. Positioning these initiatives with respect to each other shows that ODP is providing the overall framework architecture and OMG is providing the key components populating that framework.

ODP provides a framework for designing and building any kind of distributed system. Herbert [3] gives a good overview. To exploit ODP in support of their specific requirements, the telecommunications providers and vendors have formed the Telecommunications Information Networking Architecture Consortium (TINA-C) [9]. This is an industrial initiative aiming to use ODP as the basis of a common telecommunications architecture that meets the requirements of intelligent networks (IN) [10] and the telecommunications management network (TMN) [11]. The consortium plans to start field trials in two or three years' time and to deliver a validated reference implementation (and associated specifications) in five years' time.

2.3   OBJECTS IN ODP

The developing ODP standards will enable the application designer to construct applications without the need for detailed knowledge of the underlying operating system or communications infrastructure. This fundamental abstraction is captured by the ODP computational language, which allows the designer to specify and program an application as a collection of interacting objects. The mechanisms (which are also characterized as objects) that support this abstraction are defined in the ODP engineering model. These computational and engineering views of a simple request/reply interaction between a client and a server are illustrated in Fig. 2.1. Each of the ovals or boxes represents an object.

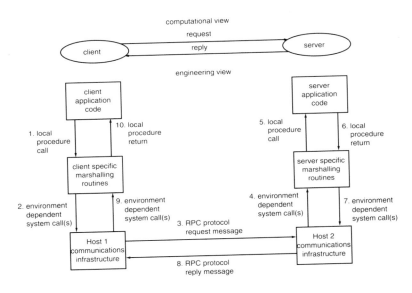

Fig. 2.1   Computational and engineering views of a simple request/reply interaction between a client and a server.

In this way, the concerns of the application designer and the system designer are separated. The application designer can concentrate on the functionality of the application; the systems designer is concerned with constructing and maintaining the engineering mechanisms that comprise the computational platform. A bridge from one model to the other can be provided in the form of tools that automatically generate the engineering code for the chosen environment. For example, the marshalling routine objects in Fig. 2.1 can be generated in this way, and the client and server application code objects represent compilations of the computational client and server objects.

2.4   TEMPLATES

In order to understand objects it is necessary to describe their properties. This is the role of ODP templates. Templates are used to describe objects in terms of the services they provide. Objects providing the same service can be described by the same template. A template describes their common features, and abstracts from their differences. Their common features might, for example, include state parameters and operations. One of the main differences abstracted is the initial state assigned to the objects when they are created (see Table 2.1 for ODP's definitions).

Unlike the templates of OSI management's GDMO notation used for describing managed objects [12], ODP templates are not forms to be filled out. Rather, an ODP template is the complete specification of a collection of objects. The template may have been produced by filling out a set of forms, or by using an appropriate language.

The process of instantiation is used to produce a new object from a template. Typically, this entails establishing initial values for any state parameters. For example a buffer object might be created with empty contents.

In summary, a template describes the common features of a collection of objects. The objects can be generated from a template by the process of instantiation. Figure 2.2 illustrates the relationship between the concepts of template, object and instantiation.

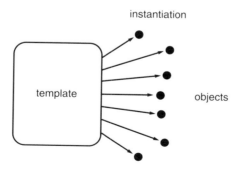

Fig. 2.2   Instantiation of a template to produce objects.

Table 2.1 ODP definitions.

Object	A model of some real world entity. An object is encapsulated, i.e. any change in its state can only occur as a result of interaction through a well-defined interface with the environment of the object.
Template	A specification of the common features of a collection of objects.
Instantiation	The process which takes a given template and results in the existence of a new object in its initial state. The new object may be called an instantiation of the template.
Type	A predicate. An object is of the type, or satisfies the type, if the predicate holds for the object.
Class	The set of all objects satisfying a type.
Subclass	One class is a subclass of another precisely when the first class is a subset of the second.
Subtype	One type is a subtype of another precisely when all objects which satisfy the first type also satisfy the second type.
Template-type (preliminary definition)	A predicate expressing that an object satisfies the properties described in a given template.
Behavioural compatibility	An object is behaviourally compatible with a second object, with respect to a fixed set of criteria, if the first object can replace the second object without the environment detecting the difference.
Instance	An object is an instance of a specified template when it is an instantiation of some template that extends the specified template.
Template-type (final definition)	A predicate expressing that an object is an instance of a template.
Incremental inheritance	The derivation of a new template (the derived template) by incrementally modifying an existing template (the parent template).

Objects instantiated from different templates may share many similarities. As a result, it is often useful to be able to classify objects independently of their templates. Types can be used for this purpose.

2.5 TYPES AND CLASSES

Many object oriented programming languages do not explicitly identify the concepts of template or type. For example, Smalltalk [13] does not talk about types, and overloads the term class to mean both a collection of objects and the template that specifies their properties. This would appear to be unnecessarily confusing as these are fundamentally different notions. This section will distinguish the concepts of type and class, both from each other and from the notion of template.

A type is simply a predicate. For example 'is red' is a type. An object satisfies a type, or is of the type, if the predicate holds for the object. Objects do not have to be very similar to satisfy the same type; they only need to possess the properties prescribed by the type. For instance a particular flag, a particular brick house and a particular sports car might be red.

Types implicitly classify objects into sets known as classes. So each type gives rise to an associated class. Specifically, a class is the set of objects satisfying a type. In set-theoretic terms, if T is a type and x is any object, then there is a class C defined:

$$C = \{x | x \text{ satisfies } T\}$$

In this case, the statements 'x is a member of C' and 'x satisfies T' are equivalent. Classes and types are different ways of expressing the same thing. Types are intentional; they define the properties that objects of the class must satisfy. Classes, on the other hand, identify a collection of objects. This may be done intentionally, using a type to describe their properties, or extensionally by enumerating the objects. It is useful to have both terms, because sometimes we start with an idea which we capture as a predicate and then determine which objects satisfy the predicate. Other times we may start with a collection of objects (i.e. a class).

Templates, as shown above, describe the common features of a collection of objects. Clearly then they are intentional rather than extensional. So how do they differ from types? The fact is that templates do not, in themselves, define a predicate. However, a template combined with some procedure for generating objects from a template (e.g. instantiation) does define a type. This is a special kind of type called a template-type. An example of a template type is 'x is an instantiation of template t'. More will be said about template types in section 2.6.

Generally, the set of objects that satisfy a type can vary with time. For example, a quadrilateral that supports an operation for moving its vertices independently, may sometimes have parallel sides, i.e. it will sometimes satisfy the predicate 'has parallel sides'. Temporal aspects of template-types are discussed in section 2.6.

2.6   SUBTYPING AND SUBCLASSING

In general, classes are not disjoint sets; different classes may contain many of the same objects. For example, a red shape might be in the class of red objects and in the class of quadrilaterals. So having placed classified objects into classes, those classes may often need to be compared.

Because classes are sets of objects, they can be compared using standard set-theoretic operators. Subsetting of classes is so important that it is given a special name — subclassing[1]:

$C1$ is a subclass of $C2$ exactly when $C1$ is a subset of C

Subclassing represents the fact that objects may simultaneously belong to a variety of classes. It can be used to organize classes into a hierarchy. For example, the class of squares is a subclass of the class of rectangles and a subclass of the class of rhombuses (every square is also a rectangle and a rhombus); and both the class of rectangles and the class of rhombuses are subclasses of the class of quadrilaterals. This is illustrated in the class hierarchy of Fig. 2.3.

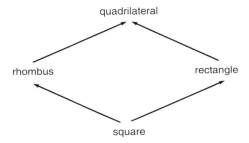

Fig. 2.3   A class hierarchy for shapes.

Related to subclassing, there is subtyping. Subtyping is implication between types. One type is a subtype of another exactly when, for all objects, satisfaction of the first type implies satisfaction of the second type:

$T1$ is a subtype of $T2$ exactly when for all x (x satisfies $T1 \Rightarrow x$ satisfies $T2$)

Subclassing and subtyping go hand in hand. Every type generates an associated class. So if there are two types $T1$ and $T2$, then there must be associated classes $C1$ and $C2$. $T1$ is a subtype of $T2$ exactly when $C1$ is a subclass of $C2$. This is easily shown from the above definitions of class, subclass and subtype. Therefore for each class hierarchy, recording subclass relationships, there is an equivalent type hierarchy recording subtype relationships. So Fig. 2.3 could be relabelled 'a type hierarchy'.

---

[1] Some programming languages use the term subclassing in a different sense. For example, in Smalltalk, a class which is derived from another class by inheritance is referred to as a subclass (of the other class). Like Stroustrup [14], ODP prefers to use the term 'derived class' for such classes. More will be said about inheritance in section 3.10.

To understand the structure of this type hierarchy, the predicates that define each of the types need to be examined. For this purpose a vector representation is introduced in this chapter — Fig. 2.4 illustrates the vector representation of a quadrilateral. A quadrilateral is described by position, which specifies the location of one of the corners, and by four vectors. Vector v_1 specifies the location of the first vertex with respect to position, v_2 describes the location of the second vertex with respect to the location of the first vertex, and so on. Equipped with this vector notation, predicates can be written down for each of the types. This is done in Fig. 2.5.

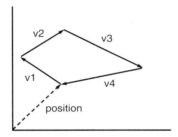

Fig. 2.4   A vector representation of a quadrilateral.

quadrilateral	v1 + v2 + v3 + v4 = 0			
rhombus	v1 + v2 + v3 + v4 = 0	v1 + v3 = 0		|v1| = |v2|
rectangle	v1 + v2 + v3 + v4 = 0	v1 + v3 = 0	v1.v2 = 0	
square	v1 + v2 + v3 + v4 = 0	v1 + v3 = 0	v1.v2 = 0	|v1| = |v2|
predicate	4-sided closed figure	opposite sides are of equal length	adjacent sides are at right angles	adjacent sides are of equal length

Fig. 2.5   Predicates describing shapes.

There are four predicates shown in Fig. 2.5. The first says that the sum of the vectors (representing the length and orientation of the sides) is zero, i.e. moving along the four sides brings you back to your original position. The second states that opposite sides are equal in length and orientation. The third invariant states that the vector product of v_1 and v_2 is 0 (the vector product of v_1 and v_2 is defined as $|v_1\|v_2|\cos\theta$, where θ is the angle between v_1 and v_2). Together with the first two invariants this means that all the angles are 90°. The fourth predicate states that adjacent sides are equal in length. Taken with the second predicate this means that all sides are of equal length.

From these predicates it should be clear that type square is stronger than (i.e. implies) both the rectangle and rhombus types, and that each of the rectangle and rhombus types are stronger than type quadrilateral — hence the hierarchy illustrated earlier in Fig. 2.3.

2.7   TEMPLATE TYPES

So far the discussion of types and classes has focused on the classification of objects according to some subset of their properties. The behaviour of objects has not yet been considered.

Consider then the shapes example in more detail. In the classification of Fig. 2.3, only values of the state parameters were considered. For example, relationships between the lengths of different sides were described. If behaviour is introduced, a more complicated picture emerges. Recall that objects are specified by templates, and consider the alternative templates (given in Fig. 2.6) for rectangle objects. The notation used here, is not from ODP and has been invented by the author for illustrative purposes only. It uses the vector notation introduced earlier.

Both rectangle1 and rectangle2 are templates describing a collection of rectangles. In the earlier discussion of types, the type rectangle essentially comprised just the invariants (and implicitly the state parameter declarations). Rectangle1 and rectangle2 additionally include operations. The semantics of these operations have not been formally defined. They are, however, illustrated in Fig. 2.7. 'Move' should be interpreted as a translation by some specified vector v. 'Stretch' is the movement of the two vertices furthest away from the origin by some specified distance, x, in the direction perpendicular to their joining line.

rectangle1	
state parameters:	v1, v2, v3, v4: vectors position: vector
invariant:	v1 + v2 + v3 + v4 = **0** v1 + v3 = **0** v1.v2 = 0
operations:	move

rectangle2	
state parameters:	v1, v2, v3, v4: vectors position: vector
invariant:	v1 + v2 + v3 + v4 = **0** v1 + v3 = **0** v1.v2 = 0
operations:	move stretch

Fig. 2.6   Alternative templates for rectangle objects.

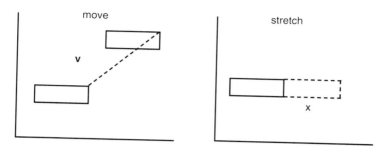

Fig. 2.7 Illustrating the move and stretch operations.

Imagine that these templates are to be used in a drawing system. Rectangle1 and rectangle2 differ in terms of the operations that they support. Rectangle1 only allows its rectangles to be moved, whereas rectangle2 allows its rectangles to be moved and to be stretched. Rectangle2 is therefore the more powerful template. It should be noted that both 'move' and 'stretch' preserve the invariant.

Now a template for squares is defined. In this case, the 'stretch' operation must be ruled out as it would invalidate the new invariant $|v_1| = |v_2|$. The template is given in Fig. 2.8.

square	
state parameters:	v1, v2, v3, v4: vectors position: vector
invariant:	v1 + v2 + v3 + v4 = **0** v1 + v3 = **0** v1.v2 = **0** $\|v1\| = \|v2\|$
operations:	move

Fig. 2.8 The square template.

Now recall that objects can be created from a template by instantiation. To instantiate the square template, initial values for the state parameters position, v_1, v_2, v_3 and v_4, must be assigned. So instantiation does not change any of the properties defined in the template. It can therefore be seen that instantiations of the square template can satisfy all the properties of rectangle1. Hence the predicate 'satisfies template square' is a subtype of the predicate 'satisfies template rectangle1'. Conversely, instantiations of square cannot satisfy all the properties of rectangle2, because squares cannot be stretched. Hence the predicate 'satisfies template square' is not a subtype of the predicate 'satisfies template rectangle2'. This is reflected in the type hierarchy of Fig. 2.9. Figure 2.9 also illustrates the fact that rectangle2 is a subtype of rectangle1 (because any instantiations of rectangle2 satisfy the invariants and support the operations of rectangle1).

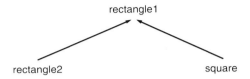

Fig. 2.9   Another type hierarchy for shapes.

There are three important points here. The first is that types can be as detailed as you want them to be. In particular, types may or may not address the behaviour of objects. The second point is that the level of detail can affect the type hierarchy. In the first analysis no attention was paid to the operations that can be performed on shapes, with the result that square was definitely a subtype of rectangle. Now it has been seen that, when behaviour is considered, square is not necessarily a subtype of rectangle. In particular, this depends on what operations rectangle supports (i.e. on whether rectangles are described by rectangle1 or rectangle2).

The third point is that a template has been seen to be used as the basis of a type definition. This kind of type is referred to as a template-type. Each template gives rise to an associated template-type, and all instantiations of a template satisfy the associated template-type for their entire lifetime.

To determine whether an object satisfies a template-type, the template from which the object was instantiated needs to be compared with the template used to define the template type. In fact, a test is needed to decide whether the object template is behaviourally compatible to the template-type template. The result of such a comparison can be very useful. If one template is behaviourally compatible to another template then it means that instantiations of the first template can be used instead of instantiations of the second template. This is important for system upgrade and for improving availability when the original service is being used. Behavioural compatibility is discussed in the next section.

Because templates include a description of the behaviour of a collection of objects, template types therefore also describe behaviour. This has a particular significance because the role of behaviour is to describe the dynamic properties of objects, i.e. it describes how properties change over an object's lifetime. This means that a template-type describes the properties of a collection of objects over their whole lifetime. In contrast to this, some types ignore behaviour, i.e. they do not describe the behaviour of objects. Such types may therefore be satisfied at certain points in the lifetime of an object but not at other points. In general, types present a partial and transitory view of the properties of objects. To obtain a lifelong classification of objects, their whole behaviour must be considered. This is the role of template-types.

2.8   BEHAVIOURAL COMPATIBILITY

Behavioural compatibility is a means of comparing the behaviour of either two objects or two templates. An object is a model of a single entity in the real world. A template is a model of a collection of entities in the real world. So because both kinds of model can be characterized as behaviours, behavioural compatibility applies equally to objects and templates.

One behaviour is said to be behaviourally compatible with another, if the first behaviour can replace the second behaviour in some environment, without the environment being able to detect any difference. Any particular interpretation of behavioural compatibility will impose constraints on the allowed behaviour of the environment. A common approach is to assume that the environment behaves as a tester for the original object. That is, the environment should be capable of fully exercising the original behaviour but should be able to do no more. Such assumptions are essential if behavioural compatibility in the context of some unknown environment is to be considered.

ODP defines two kinds of behavioural compatibility — natural and coerced. If one behaviour has to be modified before it can replace another behaviour, then there is coerced behavioural compatibility. An example of such a modification might be the hiding of additional parameters on certain interactions. In this way an interaction of the new behaviour can be made to look like an interaction of the orignal behaviour. If no modification is necessary, then there is natural behavioural compatibility. Natural behavioural compatibility is needed to determine whether an object satisfies a template-type.

In general, behavioural compatibility is reflexive but not necessarily transitive[1]. Such relations are very useful, but for determining satisfaction of template-types, a transitive relation is required — if square is behaviourally compatible with rectangle1, and rectangle1 is behaviourally compatible with quadrilateral, then square can be expected to be behaviourally compatible with quadrilateral, too.

---

[1] A relation is reflexive if every element is related to itself. A transitive relation has the property that whenever elements x and y are related, and elements y and z are related, then x and z must be related.

Within this chapter the term 'extends' is used to represent a natural, transitive behavioural compatibility relation.

2.9 INSTANTIATIONS AND INSTANCES

Earlier it was mentioned that the instantiations of a template are not the only objects to satisfy the associated template type. In recognition of this, a new term is introduced — the objects that satisfy a template-type are called 'instances' of the template. The instantiations of a template form a subset of the instances of the template.

All objects are instantiations of some template. So if a given instance of a template is not an instantiation of that template, then it must be an instantiation of another template. In such cases the second template must be behaviourally compatible to the first template. For example an instantiation of square is also an instance of rectangle1. This arises because square is behaviourally compatible to rectangle1.

An object is defined to be an instance of a specified template exactly when it is an instantiation of some template that extends the specified template.

Now that the term instance has been introduced, what it means to satisfy a template-type can be clarified. Template-type is defined as a predicate expressing that an object is an instance of a template. So an object satisfies a template-type exactly when it is an instance of the associated template.

The relationship between instantiations, instances, templates and classes is illustrated in Fig. 2.10.

Figure 2.10 shows how the instances of template t are related to the instantiations of templates s and t. The relationship is best explained by considering the instantiation o of template s. Because template s is behaviourally compatible to template t, it can be deduced from the definition of the concept of instance, that o must be an instance of t. Similarly all other instantiations of s are instances of t. Thus it can be seen that the instances of t include both the instantiations of t and the instantiations of s. Similarly it can be deduced that the instances of s would include the instantiations of s and the instantiations of any template that is behaviourally compatible to s. As a result, the behavioural compatibility relationship between templates induces a subsetting relationship between the associated classes, e.g. the set of instances of t contains the set of instances of s.

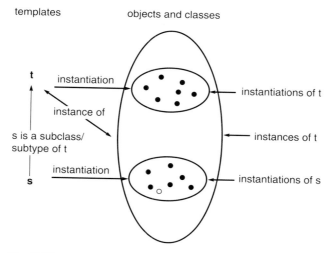

Fig. 2.10   Relationship between templates, classes and objects.

2.10   ORDERINGS ON TEMPLATES, TEMPLATE-TYPES AND TEMPLATE-CLASSES

It has already been shown that each template gives rise to an associated template-type. Similarly it generates a template-class — the set of instances of the template. If C is the template-class, t is the template and T is the template type, where T = 'is an instance of t', then:

$$C = \{x \mid x \text{ satisfies } T\} = \{x \mid x \text{ is an instance of } t\}$$

Because they are (respectively) types and classes, template-types and template-classes can be organized (respectively) into subtype and subclass hierarchies. The hierarchies are equivalent to the hierarchy that records behavioural compatibility between templates. Figure 2.11 records the correspondence between extension, subtyping and subclassing. It also identifies the formal definition of the various hierarchy orders. The formal definitions of subtyping and subclassing (namely implication and subsetting)

	templates	types	classes
ordering	extension	subtyping	subclassing
formal definition	*language specific*	implication	subsetting

Fig. 2.11   Extension, subtyping and subclassing.

are well-defined and widely understood mathematical concepts. A formal definition of extension, on the other hand, depends on the sepcification language used to describe behaviour. Readers familiar with the specification languages LOTOS or Z are referred to the definitions of extension in Brinksma and Scollo [15] and Cusack and Rafsanjani [16] respectively. These may additionally be found in part 4 of the ODP reference model. Other aspects of object oriented specification in Z are discussed in Chapter 7.

2.11   INHERITANCE AND SUBTYPING

In the wider object oriented programming community, inheritance is used with a variety of meanings, though there are perhaps two main interpretations. The first is subtyping which has already been discussed; the second is what ODP calls incremental inheritance.

Incremental inheritance is the derivation of a new template through the modification of an existing template. The new template is called the derived template and the original template is called the parent template, and by analogy the class associated with the derived template is called the derived class. The modifications can be of any kind, including adding or deleting state-parameters, adding, deleting or modifying operations, and strengthening or weakening invariants. For example, the template square could be derived from both rectangle1 and rectangle2 by incremental inheritance.

The templates, square, square' and square'' (presented in Fig. 2.12), are equivalent and describe equivalent classes of objects. All three templates are subtypes of rectangle1, but none of them are subtypes of rectangle2. The example of square'' demonstrates that subtyping and incremental inheritance are independent concepts; square'' inherits from rectangle2 but is not a subtype of rectangle2. Square' on the other hand, demonstrates that incremental inheritance and subtyping can co-exist; square' inherits from and is a subtype of rectangle1. These differences are illustrated in the inheritance and subtyping graphs of Fig. 2.13.

square'	
inherits:	rectangle1
new invariant:	$|v1| = |v2|$

square''	
inherits:	rectangle2
new invariant:	$|v1| = |v2|$
delete operations:	stretch

Fig. 2.12   The square' and square'' templates.

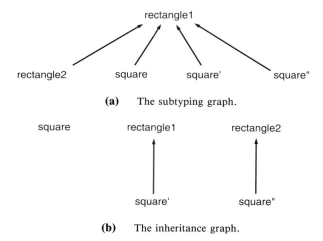

(a)   The subtyping graph.

(b)   The inheritance graph.

Fig. 2.13   A comparison of inheritance and subtyping graphs.

2.12   CONCLUSIONS

This chapter has presented in some depth a very small part of ODP — a selection of ODP's object oriented concepts — and is important on two levels.

On a technical level, it provides useful clarification of a number of concepts which are often confused. In particular the following concepts have been distinguished:

- incremental inheritance and subtyping,
- template and class,
- instantiation and instance,
- type and template-type.

On a strategic level, this chapter presents an area of ODP which has applicability beyond distributed systems. The existing plethora of object oriented terms and interpretations means that a reference set of definitions is required. Portability of skills dictates that there should be the same set of definitions for distributed object oriented systems as for non-distributed object oriented systems. After all, single machine systems are just a special case of distributed systems. Moreover there may be a need to expand a single machine system into a distributed system at some later date. The general approach taken by ODP provides a suitable common reference model.

REFERENCES

1. ISO 10746-2 RM-ODP Part 2: Descriptive Model, Ottawa (June 1992).

2. Bowen D: 'Open distributed processing', Computer Networks and ISDN Systems, 23 (1991).

3. Herbert D: 'The challenge of ODP', IFIP Transactions C-1 Open Distributed Processing, North Holland, Proceedings of the IFIP TC6/WG6.4 International Workshop on ODP, Berlin (October 1991).

4. 'The common object request broker: architecture and specification', OMG document number 91.12.1 Revision 1.1 (December 1991).

5. ISO 10746-1 RM-ODP Part 1: Overview and Guide to Use, Boulder (November 1992).

6. ISO 10746-3 RM-ODP Part 3: Prescriptive Model, Boulder (November 1992).

7. ISO 10746-4 RM-ODP Part 4: Architectural Semantics, Boulder (November 1992).

8. Network Monitor, 6, No 11, Patricia Seybold's Office Computing Group (November 1991).

9. TINA International Consortium Charter, Issue 7 (October 1992).

10. CCITT Q1200 series: 'Intelligent Networks'.

11. CCITT M3010: 'Principles for a Telecommunications Management Network'.

12. IS 10165-4 OSI Structure of Management Information: Guidelines for the Definition of Managed Objects (GDMO).

13. Goldberg A and Robson A: 'Smalltalk-80: the language and its implementation', Addison-Wesley (1983).

14. Stroustrup B: 'The C++ programming language', Addison Wesley (1991).

15. Brinksma E and Scollo G: 'Formal notions of implementation and conformance in LOTOS', Technical Memorandum (INF-86-13), University of Twente (December 1986).

16. Cusack E and Rafsanjani G-H B: 'ZEST', in Stepney S, Barden R and Cooper D (Eds): 'Object orientation in Z', Springer Verlag (1977).

3

OBJECT ORIENTED KNOWLEDGE REPRESENTATION

P M Yelland

3.1   INTRODUCTION

The work detailed in this chapter had its origins in the needs for an information storage system or repository for use by applications managing telecommunications networks. It was the intention to construct a repository that would be applicable throughout the entire gamut of business activities. This meant that it might contain large volumes of relatively simple data relating to the performance of a communications network, while at the same time storing smaller amounts of highly complex information concerning the types of service provided by that network. Support for the storage and manipulation of such disparate kinds of information was to be one of the primary features of the repository.

   This chapter describes initial attempts to produce an information base meeting these requirements. It involved extending the Smalltalk object oriented programming system [1] with 'term classification' facilities like those found in the knowledge representation language KL-One [2] and its successors. In the next section, we examine the reasons for choosing this approach. Subsequent sections describe the term classification facilities offered by the system and relate the possible repercussions of our investigations for object oriented software development in telecommunications and other areas. We conclude by indicating possible directions for future research.

3.2   APPROACHING OBJECT ORIENTED KNOWLEDGE REPRESENTATION

As we stated at the outset, one of our overriding concerns was to address the problem of providing storage and manipulation facilities both for large amounts of simple data and smaller amounts of complex information. It was our opinion that in order to satisfy these requirements adequately, our prospective repositories would need to incorporate at least two different mechanisms for the representation and storage of information. One might consider, for example, using a traditional database for simple, voluminous data, and an expert system for smaller amounts of more complex information. Each component of such a system would be inadequate in isolation — representing complex information in a traditional database is often an involved and cumbersome affair, and few expert systems have the performance characteristics necessary to deal efficiently with large bodies of data — but their combination might meet our objectives. In fact, we did produce such a system in the early stages of our investigations (see Bujnowski et al [3] for further details).

However, experience quickly demonstrated the shortcomings of such an approach — aside from the patently unwieldy nature of a storage system with two large independent components (i.e. database and expert system), it became apparent that the wide divergence in the representation schemes used by each component was a major impediment when using the system. This dissimilarity obstructed the communication of information between the components of the repository, led to premature 'freezing' of design decisions regarding the proper location of information[1], and greatly complicated maintenance procedures.

This experience led us to seek to reduce the disparity between the representation schemes used by the different storage technologies in our repositories; ideally, there would be little or no apparent discontinuity between them. To this end, we elected to use an object oriented framework to incorporate the different representation mechanisms behind uniform object interfaces. It is also widely held that the object oriented approach is well-suited to the modelling of the kind of information encountered in the

---

[1]Should, for example, information relating to the operational status of a network exchange be located in the database, in the expert system, or in both?

management of large-scale communications networks. In addition, this decision provided a ready means of meeting one of the requirements on our repositories — selecting a mature object oriented programming system as the basis for our implementation would give us access to the kind of facilities needed to handle the voluminous, simple information alluded to above[1].

However, we felt that the actual programming notations offered by many object oriented programming systems were inadequate for encoding the more complicated forms of information which we intended to store in our prospective information repositories. Many authors (Genesereth and Nilsson [4], for example) have averred that representing complex information using low-level procedural code results in a loss of clarity which can make a system difficult to implement, extend and maintain. What we wanted was a high-level representation mechanism. Such high-level mechanisms for representing complex information are sometimes referred to as knowledge representation schemes.

When we came to examine previous attempts to integrate knowledge representation and object oriented programming, we decided that they could be broadly categorized either as rule-based programming facilities for object oriented systems, or as combinations of logic and object oriented programming (a comprehensive survey of both types of system may be found in Saunders [5]). We have deliberately omitted discussion of frame-based systems because, for our purposes, their knowledge representation capabilities do not differ significantly from those of object oriented programming systems.

Unfortunately, we felt that both of these approaches were at odds with our desire for a 'seamless' combination of representation schemes in our information repositories. For example, the control flow in a rule-based programming system is quite different from that in an object oriented one; whereas control is normally passed explicitly in an object oriented system by the dispatch of messages, in a rule-based environment, control is seized when certain patterns emerge in a shared 'working memory'. In addition, whereas objects persistently encapsulate the state of a system, the state associated with rules is normally bound only transiently and shared globally through the working memory. The mechanics of most logic programming languages are even more alien to mainstream object oriented systems — contrast, for example, Prolog's unification-based parameter-passing and backtracking control flow with the more conventional operation of C++ or Smalltalk. Such considerations prompted us to look for other ways of representing knowledge in object oriented systems.

---

[1] Such information could be encoded simply as objects in the underlying system.

3.3   KL-ONE AND TERM CLASSIFICATION

As was suggested in the introduction, the techniques we actually decided to use in our information stores originated with the KL-One knowledge representation system. These techniques — described collectively as **term classification** — have been studied extensively since the introduction of KL-One. They are introduced briefly in the next section.

3.3.1   What is term classification?

Space limitations forbid an extensive discussion of the whole range of term classification techniques — a comprehensive summary of work in this area may be found in Sowa [6]. The following gives a fairly terse introduction to the subject; more detailed definitions are presented in the remainder of the chapter as they are encountered.

As with object oriented programming, individual capsules of information — objects — form the basis of term classification. In term classification, an object is an individual entity with a number of externally visible attributes, known as roles. These roles may contain a single value (the time-in-service of a network component might be one such), or a collection of values (the voice-circuits of an exchange, for example)[1]. Unlike the objects in object oriented programming systems, objects in term classification have no 'invisible' attributes or private variables, nor do they respond to messages. However, most term classification systems allow objects to be partially specified, rather than insisting on a complete definition, as do most object oriented programming systems. For example, in a term classification system, it is possible to state that the time-in-service of exchange-017 is greater than 25 days (making it eligible for imminent inspection, say), without having to state precisely how long it has actually been in service.

Term classification systems also feature entities — known as concepts — which are analogous to classes in object oriented programming. However, unlike classes, which describe the internal structure and behaviour of their instances, concepts simply represent collections of objects with certain properties. The objects comprising a concept are described by a term that lays out a number of conditions which the roles of those objects should satisfy. For example, we could describe the concept HighlyConnectedComponent with a term specifying objects with more than 100 voice-circuits (i.e. one hundred or more objects in their voice-connections roles); members of the concept

---

[1] The actual definitions of network entities used in this paper are purely illustrative, and not intended to represent elements of a realistic network model.

ReliableComponent might be objects whose time-in-service was greater than 300 days. Furthermore, term classification allows concepts to be combined, 'intersecting' their term descriptions to form a new one. So, for example, the concept ReliableHighlyConnectedComponent may be defined as the combination of the concepts ReliableComponent and HighlyConnectedComponent; its members lie in the intersection of these two concepts, being objects with one hundred or more voice-circuits whose times-in-service are greater than 300.

3.3.2   Subsumption and classification

Term classification systems allow hierarchies of components to be constructed, rather like the inheritance hierarchies of classes in object oriented systems. The hierarchies constructed in term classification, however, are more akin to the type hierarchies discussed in Chapter 2 than the inheritance hierarchies found in most object languages[1].

Most of the time, the assertion that one class in an object oriented system is a subclass of another implies some sort of set/subset relationship. So, for example, the assertion that the class Circle is a subclass of GeometricObject tends to accord with the expectation that circles are geometric objects. In general, however, (typing considerations aside) this set/subset relationship has no formal effective status; the system will not try to ensure that Circles 'conform' in some sense to the notion of GeometricObject (see America [7] for proposals to enhance object oriented programming languages along these lines). Indeed, there are occasions when a sub/superclass relationship is used purely as a means of gaining access to code in another class. (A prime example of this is the class Dictionary in the Smalltalk system, which is a subclass of class Set, even though the behaviour of Sets and Dictionaries differs quite markedly.)

By contrast, in term classification systems, where concepts are simply collections of objects, the concept/subconcept relationship mirrors exactly the set/subset relationship. So the concept ReliableHighlyConnectedComponent defined in the previous section is a subconcept of ReliableComponent because by definition, all ReliableHighlyConnectedComponents must be Reliable-Components too. In fact, by examining the terms which define concepts, it is possible to derive such relationships automatically. Not all of these relationships need be immediately apparent from a concept's definition. For example, the concept ReliableHighlyConnectedComponent may well be a subclass of VoiceCarrier (defined as a component with at least one voice-

---

[1] In fact, much of the discussion in this section parallels that in Chapter 2; *concepts*, for example, are akin to Rudkin's *classes; terms* are *types; subsumption* may be read as the *subtype* relationship, and so on. In this chapter, we use established vocabulary used by workers in term classification, for the sake of consistency.

circuit), even though the concept VoiceCarrier is mentioned nowhere in the definition of ReliableHighlyConnectedComponent. Detecting concept/sub-concept relationships of this kind (known in the vernacular as subsumption relationships) is a central concern of many term classification systems.

The detection of subsumption relationships — both explicit and implicit — allows a term classification system automatically to construct a hierarchy of concepts, into which newly defined concepts may be placed. The process of building such hierarchies is known as **classification** (and since it is achieved by examining the terms which define classes, it gives rise to the epithet 'term classification').

3.3.3 Combining term classification and object oriented programming

Of all the approaches to knowledge representation which we reviewed, term classification, based as it is on objects and classes, appeared to be the most natural candidate for amalgamation with object oriented programming. In this section, we take a brief look at previous exercises in this area, and give an overview of our own combined system.

3.3.3.1 Previous work

The degree of commonality between term classification and object oriented languages has already been noted by other researchers in the field — most notably in Patel-Schneider [8]. In addition, there have been projects in the past which have sought to combine term classification with object oriented programming, at least to some degree. KL-One itself, for example, made provision for attaching procedures and data to items of information (see Brachman and Schmolze [2]). KloneTalk — an implementation of KL-One in an early version of Smalltalk — is described in Fikes [9]. The 'Boolean classes' of McAllester and Zabih [10] build a restricted form of term classification into object oriented programming languages. There are also systems like Login [11], which unify term classification and other types of programming languages. To our knowledge, however, there have been no attempts at a comprehensive integration of term classification and object oriented programming on the scale of that described in this chapter.

3.3.3.2 A hybrid system

We began the constuction of our system by selecting an existing object oriented programming system and supplementing it with term classification facilities. For this purpose, we chose Smalltalk[1], primarily because of the accessibility of fundamental components in the Smalltalk programming environment (for example, in Smalltalk, the behaviour of classes themselves can be altered by the programmer).

The next part of this chapter describes the resulting system in detail. Briefly, we began by implementing a special form of Smalltalk class which we called the Concept. Concepts in our system have the functionality both of Smalltalk classes and of concepts in the term classification sense; similarly, their instances are native Smalltalk objects with the capabilities of objects in a term classification system. Concepts and their instances are pivotal in our knowledge representation scheme — their dual nature allows the subsumption and classification mechanisms of term classification to be combined with capabilities such as message-sending and inheritance offered by the host object oriented programming system.

Like concepts in a term classification system, Concepts describe their instances using terms. Therefore, unlike conventional Smalltalk classes, whose instances are determined explicitly by the programmer when those instances are created, the instances of Concepts can be determined automatically by comparing existing objects with the descriptions given by Concepts. This has two main consequences:

- since an object may match the descriptions of more than one Concept simultaneously, it may have more than one class[2] — additionally, because the instance descriptions of Concepts may specify aspects of an object's state, a change in the state of an object may lead to revision of its set of classes;

- some of the sub- and superclasses of Concepts can be determined automatically by detecting subsumption relationships. This is because the system may conclude that a Concept c_1 is a sub-class of Concept c_2 if c_1's instance description describes a subset of those objects described by the instance description of c_2. (It should be noted that, in general, this may entail some form of multiple inheritance.)

The next sections outline in more concrete terms the central characteristics of the classification facilities (from now on normally referred to collectively as 'the classifier') in our hybrid system — i.e. the definition of Concepts and the manipulation of their instances.

---

[1] More precisely, ObjectWorks® \Smalltalk™ (Release 4) from ParcPlace Systems, Inc.

[2] Remember that Concepts are merely special classes, so we use the word 'class' (and 'sub-' and 'superclass') to describe both Concepts and regular Smalltalk classes. Incidentally, multiple class membership (objects with more than one class) is a feature of some object oriented programming systems without term subsumption facilities [12].

3.4   TERM CLASSIFICATION FACILITIES IN THE HYBRID SYSTEM

3.4.1   Objects, roles and role restrictions

Recall that objects appear to the classifier as individuals which possess a number of roles. In general, these roles are sets containing other objects. (Since instances of Concepts are also Smalltalk objects, they may possess instance variables in addition to their roles. The roles of an object are distinct from its instance variables, which are of no interest to the classifier.) Concepts describe their instances by specifying a number of restrictions (comprising the 'terms' mentioned in the previous section) which their instances' roles must satisfy. In the following, we will show how roles and role restrictions are constructed, how restrictions can be combined, and how they are used in forming Concepts.

3.4.1.1   Defining roles

Two types of role may be associated with an object. **Primitive** roles are sets whose contents are determined explicitly by adding or removing elements. The contents of **defined** roles, on the other hand, are derived implicitly from the contents of other roles of the object. This is achieved in general by forming the intersection of a set of roles, and then selecting those elements of the intersection which are members of given collections or instances of given classes. Roles are usually given global names, and instances of more than one Concept may have roles with the same name.

As an example of role-definition, imagine that in modelling some part of a communications network we decide to declare a primitive role Circuits, intended to record the connections available to a given element of the network. We could do this by evaluating the following expression:

```
PrimativeRole name:#Circuits
        category: 'Elements-Roles'
```

The specification of a category is used to determine how the role is displayed by system browsers; we will omit categories in future declarations.

Next, assuming that the Concept LiveConnection describes objects representing viable connections in a network, and that the role TrunkConnections holds all of the trunk connections of a network element, we can declare a defined role which contains all those circuits of a network element which are also live trunk connections:

```
DefinedRole name:#LiveTrunkCircuits
      subroleOf: Circuits, TrunkConnections
      restrictions:LiveElement
```

3.4.1.2   Role restrictions

The experimental classifier recognizes role restrictions of two sorts:

- **cardinality** restrictions limit the size of an object's roles;
- **value** restrictions constrain the types of object which may occur in the roles of an object.

Cardinality restrictions restrict the size of a role by imposing a lower and/or an upper bound on it. For example, an object with at least 30 circuits would satisfy the restriction:

```
Circuits atleast:30
```

and one with at most 40 circuits would satisfy:

```
Circuits atmost:40
```

Value restrictions compel the objects in a role to be instances of a given Smalltalk class (or one of its subclasses), instances of a Concept (or one of its subclasses), or members of a Smalltalk Collection. For example, assuming that an object has a role State, we might wish to have its state described by the tokens #operational and #nonOperational:

```
State all: (Set
             with:#operational
             with:#nonOperational)
```

so that the role restriction above could be expressed:

```
State all: OperationalState
```

Alternatively, we might want a rather broader description of state which could be any Smalltalk String:

```
State all: String
```

We could also specify that all circuits of an object are members of the Concept LiveElement (or one of its subclasses):

Circuits all: LiveElement

3.4.1.3   Compound role restrictions

Role restrictions are combined by conjoining them with the ' , ' (comma) operator. Thus an object with between 30 and 40 circuits, all of which were live, would satisfy the conjunction:

(Circuits atleast:30),
(Circuits atmost:40),
(Circuits all:LiveElement)

The system provides a simple 'macro' facility to allow restrictions to be combined conveniently. For example, one could define:

r between: l and: u $\equiv$ (r atleast: l), r atmost: u)
r exactly: n $\equiv$ r between: n and: n
$r = e$ $\equiv$ (r all: (Set with: e)),
(r exactly: 1)
r memberOf: c $\equiv$ (r exactly: 1), (r all: c)

3.4.2   Concepts

A Concept is defined by giving a set of classes which must be among its superclasses and a set of restrictions which its instances must satisfy. By consolidating its restrictions with the descriptions of its explicitly-specified superclasses, the classifier arrives at a comprehensive description of the Concept's instances which can then be used to find other superclasses and subclasses. As with roles, the classifier distinguishes primitive and defined Concepts.

Primitive Concepts give a description of their instances which is accurate but not necessarily exhaustive. That is to say that while every instance of a primitive Concept must satisfy its instance description, not every object which satisfies its description is necessarily one of its instances. One implication of this is that while the system can automatically determine the superclasses of a primitive Concept, it must be told explicitly if a Concept is to be a subclass of a primitive Concept. Another implication is that the system must also be expressly informed if a primitive Concept is to be one of the classes of an object or a superclass of one of those classes.[1] In our system, primitive Concepts may define instance-variables for their instances, just like Smalltalk classes. As an example, a declaration of the primitive Concept NetworkElement is given below; it is a subclass of Object, and each of its instances has a single IntegerIdentifier, a single State which is a member of OperationalState, and a private instance variable named 'comment':

```
PrimitiveConcept name: #NetworkElement
   subclassOf: Object
   restrictions: (Identifier memberOf:Integer),
           (State memberOf: OperationalState)
   instanceVariableNames: 'comment'
```

In contrast to primitive Concepts, the descriptions of **defined** Concepts describe their instances precisely; every instance of a defined Concept must satisfy its description, and every object which satisfies its description is an instance of it. This means that the system is able to deduce both the sub- and superclasses of a defined Concept, and is able to conclude that a given object is one of its instances without being told so expressly. The experimental classifier does not permit a defined Concept to declare any additional variables for its instances; all must be inherited from its superclasses. (In conjunction with the restrictions on class changes described below, this structure obviates the need to restructure an object each time its set of classes is altered.) The Concept LiveElement referred to above might be defined simply as a subclass of NetworkElement, all of whose instances have a State which is #operational:

```
DefinedConcept name: #LiveElement
   subclassOf: NetworkElement
   restrictions: (State = #operational)
```

3.5   USING OBJECTS WITH THE CLASSIFIER

3.5.1   Creating objects

Instances of Concepts are created — just like instances of any other Smalltalk class — by sending a message (generally new) to the Concept in question (it should be recalled that as their state changes, objects may later be found to be instances of classes other than their creator). Complications arise from

---

[1] Henceforth, we abbreviate 'superclass of a class of an object' to 'superclass of an object'.

the dependence of an object's classes on its state. For example, it is possible that some oversight by the programmer might lead to objects newly created by the defined Concept LiveElement declared above having a State which was not equal to #operational and thus did not satisfy the instance description of the Concept. This might lead to class LiveElement somewhat paradoxically producing new objects which were not legally its instances. At the moment, the default implementation of the message new in Concepts takes steps to detect such anomalies by sending the message initialize to the newly created objects (the method for initialize is supplied by the creator of the Concept), and then finding their classes on the basis of their resulting state. Should the Concept not be amongst the classes of a new object after this initial classification, an exception is raised.

3.5.2   Changing the classes of an object

Once an instance of a Concept has been created, there are constraints on the way in which its set of classes may change throughout its life. Recall from the discussion in the previous section that the description which a primitive Concept gives for its instances is not complete. We noted that this implied that the classifier had to be told explicitly if a primitive Concept was to be one of the classes (or superclasses) of an object. In the actual system, there is no way of explicitly changing the classes of an object once created — all class changes must occur as a side-effect of changes in the object's state. Thus a primitive Concept can only be made a (super)class of an object by creating the object as an instance of it (or one of its subclasses) in the first place.

The classification system also enforces the converse of this restriction — once a primitive Concept has been made a class or superclass of an object, it must remain amongst that object's classes or superclasses throughout its lifetime. If the object changes so that it no longer satisfies the instance description of the primitive Concept, the system raises an exception. This allows the restrictions in the description of a primitive Concept to be used as integrity constraints for its instances. If, for example, an instance of the primitive Concept NetworkElement defined in the previous section were to try to record a State which was not an element of OperationalState, then it would cease satisfying the restrictions imposed by that Concept's description. Since the system cannot remove NetworkElement from the object's classes, it raises an exception instead, indicating a violation of the constraint. The classifier can be used in this way as a form of run-time type system for objects, as described later in this chapter.

3.5.3   Modifying objects

As far as the classifier is concerned, all modifications of objects occur as a result of altering the contents of roles (changes to instance variables go unnoticed by the classifier). Roles implement the protocol of Smalltalk's abstract class Collection, so their contents may be altered using standard messages, such as add: or remove:. A role is retrieved simply by sending its owner a message consisting of its name with the initial letter in lower case (e.g. aNetworkElement state). Classification of an object (i.e. adjusting its classes to reflect its current state) is invoked explicitly by sending it the message classify. This permits complex role modifications to take place without incurring the overhead of classification at each step, and also allows an object temporarily to violate integrity constraints between classifications.

The system automatically maintains the relationships between an object's primitive and defined roles determined by their definitions. For example, recall the definition of the defined role LiveTrunkCircuits:

```
DefinedRole name: #LiveTrunkCircuits
       subroleOf: Circuits, TrunkConnections
       restrictions:LiveElement
```

Executing the statement 'anExchange live TrunkCircuits add: aCircuit', the system will first verify that the element added to the role (aCircuit) is actually live (i.e. that LiveElement is one of its classes or superclasses) — if not, an exception will be raised. The element is then added to role LiveTrunk Circuits, and since the latter is defined as the intersection of Circuits and TrunkConnections, the element is added to these roles too. Conversely, if a LiveElement is removed from Circuits or TrunkConnections, the system ensures that it is also removed from LiveTrunkCircuits.

3.5.4   Automatic classification

It is possible for an object to be classified automatically by the system, as a result of the classification of elements of its roles. To see how this might come about, consider the following definitions, which supplement the declarations of NetworkElement, Circuits and LiveCircuits given above:

```
DefinedConcept name: #DeadElement
       subclassOf: NetworkElement
       restrictions: (State = #nonOperational)
```

DefinedRole name: #DeadCircuits
    subclassOf: Circuits
    restrictions: DeadElement

PrimitiveConcept name: #Exchange
    subclassOf: NetworkElement
    restrictions: (Circuits all:NetworkElement)
    instanceVariableNames:' '

DefinedConcept: #OperationalExchange
    subclassOf: Exchange
    restrictions: (Circuits all:LiveElement)

DefinedConcept
    name: #PartiallyOperationalExchange
    subclassOf: Exchange
    restrictions: (DeadCircuits atleast: 1)

These declarations introduce the Concept of an Exchange (an object with Circuits which are NetworkElements), with subclasses OperationalExchange (all of whose Circuits are in an operational state) and PartiallyOperationalExchange (with at least one non-operational circuit).

Imagine the situation depicted in Fig. 3.1. Here we have an OperationalExchange object with one circuit — a LiveElement whose State is #operational. Now change the State of the LiveElement to #nonOperational and reclassify. This changes the class of the element to DeadElement. At this point, the system observes that the exchange object no longer qualifies as an OperationalExchange; on the contrary, it is now a PartiallyOperationalExchange. The system alters its class accordingly, with the result illustrated in Fig. 3.2.

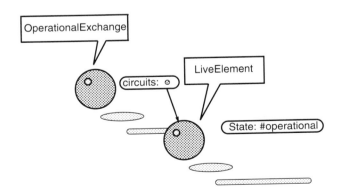

Fig. 3.1   Initial configuration.

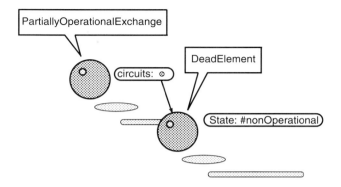

Fig. 3.2   After reclassification.

Since the change in the class of the exchange is carried out at the instigation of the system, and not of the programmer, the object is notified of its re-classification. At present, this involves sending it the message classifiedDueTo: after reclassification, the single parameter is the element whose modification caused the reclassification. As we will detail later, by supplying suitable methods for this message in Concepts, a form of 'data-driven' or 'forward-changing' inference [13] can be implemented, with the consequences of modifications to objects being propagated through the system automatically.

3.6.   CONCEPTS AND INHERITANCE

The version of Smalltalk chosen to host our prototype system does not support multiple inheritance [14]. Unfortunately, the need for such a capability arises quite naturally from the combination of term classification and object oriented programming. A user of our prototype is normally unaware of all of the sub/superclass relationships which exist between Concepts, and there is no *a priori* reason why the classification process should not uncover more than one superclass for a given Concept. In fact multiple inheritance is avoided neither in the applications produced with our own system, nor in similar ones recounted in the literature. Moreover, once devised, a suitable multiple inheritance mechanism can also be used to deal with objects which possess more than one class (again a natural product of classification), since an object with classes $c_1...c_n$ can be viewed as an instance of a postulated class c, which inherits (multiply) from $c_1...c_n$ (see Stein [15] for more details). In the light of this, therefore, it was necessary for us to devise our own multiple inheritance facilities.

Of course, one of the primary issues which must be faced in the implementation of multiple inheritance is that of resolving conflicts between attribute values inherited from more than one source. Some existing object oriented programming environments with multiple inheritance, such as CLOS [16], resolve all inheritance conflicts, using heuristics based upon the order in which the programmer declares the superclasses of classes. Unfortunately, a term classification system computes a proportion of the superclasses of a Concept automatically, largely depriving the programmer of the opportunity to supply this sort of information. One response to this might be to adopt the approach employed in extended Smalltalk [17] and Trellis/OWL [18], which requires that the user resolve all inheritance conflicts manually. However, prior experience with extended Smalltalk gave us to believe that this would be unduly burdensome. Our current prototype, therefore, uses Touretzky's inferential distance heuristic [19] (also expressed in the C++ multiple inheritance system as the rule of dominance [20]) for conflict resolution.[1] Since this does not resolve all inheritance conflicts, any which remain are signalled to the user using the system browsers, as in Extended Smalltalk.

We have made no attempt to address other issues, such as the aliasing effects which result from the 'graph-based' implementation of multiple inheritance we have chosen to use (see Snyder [14] for more details).

3.7   USING A TERM CLASSIFIER IN OBJECT ORIENTED PROGRAMMING

Our experience in using the classifier with Smalltalk led us to conclude that its scope of application exceeded that of the information bases which were our original objective. In fact, we were encouraged to find that the use of Concepts and their instances allows features of the term classifier to complement those offered by the object oriented programming system, producing a range of new facilities offered by neither component in isolation. Some of these are described in greater detail below.

---

[1] Roughly paraphrased, the inferential distance heuristic states that if a class C might inherit conflicting attributes from superclass A and B, and A is a subclass of B, then the attribute specified by A should be chosen for C. We believe that a similar heuristic was used to resolve conflicts in a multiple-inheritance system for Smalltalk-80 produced at Tektronix.

3.7.1   Assisting 'white-box' analysis

Terminological facilities are particularly useful in early so-called 'white-box' stages of most formal object oriented analysis procedures, where primary concepts in the application domain are defined. In general, these initial definitions are couched in terms of the concepts' essential qualities, rather than as 'black-box' behavioural characterizations (these normally follow in later stages of analysis).

Embodying white-box definitions as formal Concepts enables the inference procedures offered by the term classifier to be used to guide and verify the analysis. They may be used, for example, to identify incoherent (i.e. self-contradictory) definitions, or to show that two apparently distinct definitions are actually referring to the same concept.

The term classification facilities also allow the results (in the shape of Concept definitions) to be retained throughout the development of a system. Concepts allow such definitions to play an effective role in the system's implementation, where in conventional development they are normally reduced to informal annotations. In addition, Concepts can be used to make relationships between objects explicit; such relationships are often identified in object oriented analysis and design procedures, but usually represented implicitly (and rather obscurely) by chains of pointers in conventional implementations.

3.7.2   Enhanced object/class relationships

We found that one significant benefit of incorporating term classification into an object oriented programming system like Smalltalk was the introduction of the more flexible and general notion of class/instance relationship that term classification (with its preoccupation with expressive power, rather than computational efficiency) has evolved.

Recall that the term classifier views class/instance relationships simply in terms of set membership. As we pointed out earlier on, this means that generally an object in our hybrid system is not restricted to a single class throughout its lifetime nor to one class at a time. This sort of generality has often been seen as desirable in object oriented programming systems; some authors [15, 21, 22] have delineated mechanisms for ascribing multiple classes to objects and for altering classes dynamically. Such features are also exhibited by a number of more modern object oriented analysis and design methods.

3.7.3   State descriptions

The terms which constitute the instance descriptions of Concepts can allude to aspects of objects which are state-dependent (an object which satisfies the restriction 'time-in-service > 6', for example, will cease to satisfy it if its time-in-service attribute is assigned the value 5). Thus a programmer can arrange for the classes of an object to change in a controlled fashion according to state-changes, effectively using Concepts to describe object states.

The term classifier can help validate descriptions of states given in this way by ensuring that, for example, they do not overlap. The classifier will also ensure that all the states associated with a particular class of objects are propagated among classes derived from that class; for example, a Concept such as Exchange might acquire states Operational and NonOperational from its superclass NetworkComponent. Again, Concepts used as state descriptions allow constructs produced in analysis and design to be manifested effectively in a system's implementation. The fact that Concepts describing states appear in the system's class hierarchies also helps span the discontinuity between dynamic and static representations which is a common criticism of current methodologies.

3.7.4   Controlled inheritance

The ability to construct hierarchies of classes automatically from their descriptions is a central feature of most pure term classification systems. In our system, this is coupled with a conventional object oriented multiple inheritance mechanism, so that Concepts inherit methods and instance variables according to the positions in the hierarchy computed by the classifier from their descriptions. Since the hierarchy constructed by the classifier is essentially a representation of subtype relationships, the combined mechanism can be used to ensure that 'inheritance follows subtyping' in a way which is normally considered desirable in object oriented systems [12]. It may also be used to support more advanced methods of controlling (especially multiple) inheritance, such as the use of the classifiers described in Hamer [23].

3.7.5   Partial object descriptions

In many applications, particularly those characterized as 'knowledge-based' (e.g. Frontini et al [24]), objects begin life with fairly vague descriptions, which are refined as processing continues. In such applications, information completely specifying an object may not be available throughout much of its lifetime. This is at odds with the normal assumption of many object oriented programming systems, which tend to assume that objects are more or less completely specified at all times. This in turn can lead to problems when implementing knowledge-based applications using conventional object oriented systems [12].

With term classification facilities, partial object specifications are much more tractable. Postulating that object Ex-010, say, is an (otherwise unspecified) instance of class Exchange implies that it satisfies the instance description of class Exchange without requiring any further specification of its attributes. Assuming, for example, that Concept Exchange's description includes the term '(voice-circuits all: VoiceCircuit)', one could conclude, purely from the assertion that Ex-010 is an Exchange, that Ex-010's voice channels are all VoiceCircuits. Once such an assertion has been made, the classifier may be used to guarantee the validity of such a description as it is refined, by detecting incoherence, for example, resulting from further assertions (Ex-010 is unlikely to be a ServiceRecord as well as an Exchange), or by detecting 'type violations' arising from alterations to attributes or relationships, as described below.

3.7.6   Types and assertions

According to the term classifier, objects are members of classes if, and only if, they satisfy the instance descriptions of those classes. When an object ceases to satisfy the instance restriction of one of its current classes, the system may either seek out another class whose description it does satisfy, or it may raise an exception. The first alternative gives rise to the pattern-matching/rule-based mode of operation described in the next section. The second alternative employs the classifier as a form of assertion-checker or run-time type system (similar to those found in some frame-based systems like KEE [25]).

3.7.7   Rule-based programming

As we pointed out above, when the attributes of an object are altered, it is possible to have the system search automatically for classes whose instance descriptions that particular object currently satisfies. Recall also that in conventional object oriented systems late binding of method names means that an object's response to messages depends upon its current class(es). Allying the two features produces a form of 'object oriented rule-based programming'. Here a rule is represented by a Concept, which encodes the 'head' of the rule in its instance description. The 'body' of the rule is represented as method in that Concept with a fixed name such as invoke. Objects — in their capacity as elements of 'working memory' — are matched with the pattern expressed in the Concept descriptions when the classifier discovers a class/instance relationship. Dynamic binding ensures that the correct rule-body is located when the object is sent the message invoke. One feature of this scheme is that potential matching conflicts may be resolved using conventional inheritance conflict resolution mechanisms, rather than the more specialized *ad hoc* means found in production-rule systems such as OPS5 [26].

3.7.8   Object retrieval

The ability of the classification system to match classes and their instances may also be combined with Smalltalk's ability to enumerate the members of particular classes. This creates a sort of 'object-based query mechanism' like that described in Patel-Schneider [27]. From the user's point of view, a query is expressed as the instance-description of a new Concept, whose instances are determined automatically from those objects in the system. Enumerating those instances then returns the objects which satisfy the query.

3.8   LIMITATIONS AND FUTURE RESEARCH

Our investigations into the combination of term classification and object oriented programming are ongoing; we are applying our prototype system to a number of problems in the telecommunications domain, and we expect it to change (possibly significantly) to accommodate the demands of applications. There are a number of areas in which further development looks especially appealing:

- the implementation of more comprehensive, general term classification facilities — these might include those found in existing systems like Classic, or Loom, or less 'traditional' capabilities such as are suggested by Doyle and Patil [28];

- the implementation of 'specialized reasoning modules', like those described in Litman and Devanbu [29] or MacGregor and Bates [30];

- further examination (not touched on directly in this chapter) of the completeness of the classifier used in the system — our current classifier, while sound, does not discover all the subsumption relationships which might be implied by a Tarksi-style model of Concepts (see Nebel [31] for a full discussion);

- embellishments to the system's user interface — at present, the interface to the classifier consists largely of standard Smalltalk tools, with a few minor extensions for displaying Concept hierarchies.

3.9 CONCLUSIONS

We have described a modest extension to Smalltalk which we believe may be useful in expressing some kinds of complex information in a reasonably clear and readily comprehended manner. Our approach to constructing this system has been to take ideas from the field of term classification, integrating them carefully with the Smalltalk system. We felt that this approach would afford particular flexibility in balancing the relative efficiency of conventional object oriented programming against the Conceptual clarity offered by knowledge representation facilities.

Initial experience tends to suggest that on the whole, the classification system substantiates many of our initial hopes. This, and the fact that researchers [12, 15] have previously expressed the need for facilities in object oriented environments which the classifier offers (at least to some degree), confirms our belief that classification systems like that described here might make a contribution to the development of object oriented programming.

REFERENCES

1. Goldberg A and Robson D: 'Smalltalk: the language and its implementation', Addison-Wesley (1983).

2. Brachman R and Schmolze J: 'An overview of the KL-One knowledge representation system', Cognitive Science (1985).

3. Bujnowski J, Salman A and Yelland P: 'Large knowledge base systems', Internal BT Report (1992).

4. Genesereth M and Nilsson N: 'Logical foundations of artificial intelligence', Morgan Kaufmann (1987).

5. Saunders J: 'A survey of object oriented programming languages', Journal of Object Oriented Programming, 1 , No 6 (1989).

6.   'The evolving technology of classification-based knowledge representation systems', in Sowa J (Ed): 'Principles of semantic networks', Morgan Kaufmann (1986).

7.   America P: 'A behavioural approach to subtyping', Proc Workshop on the Foundations of Object Oriented Languages, Springer-Verlag (1990).

8.   Patel-Schneider P: 'Practical, object-based knowledge representation for knowledge-based systems', Information Systems, 15 , No 10 (1990).

9.   Fikes R: 'Highlights from KloneTalk', in Schmolze J and Brachman R (Eds), 'Proc 1981, KL-One Workshop', Rep 4842, Bolt, Beranek and Newman (1982).

10.   McAllester D and Zabih R: 'Boolean classes', Proc Conf on Object Oriented Programming Systems, Languages and Applications, ACM (1986).

11.   Ait-Kaci H and Nasr R: 'Login: a logic programming language with built-in inheritance', Journal of Logic Programming, 3 (1986).

12.   Cordingley E: 'Analysing texts for knowledge-based systems', in Binch-Capon T (Ed): 'Knowledge-Based Systems and Legal Applications', Academic Press (1991).

13.   MacGregor R: 'A deductive pattern matcher', Proc 7th National Conference on Artificial Intelligence, AAAI (1988).

14.   Snyder A: 'Inheritance in object oriented programming in languages', in Leuzerim M, Nardi D and Simi M (Eds): 'Inheritance in hierarchies in knowledge representation and programming languages', Wiley (1991).

15.   Stein L: 'A unified methodology for object oriented programming', in Leuzerim M, Nardi D and Simi M (Eds): 'Inheritance hierarchies in knowledge representation and programming languages', Wiley (1991).

16.   Keene S: 'Object oriented programming in Common Lisp', Addison-Wesley (1989).

17.   Borning A and Ingalls D: 'Multiple inheritance in Smalltalk-80', Proc First National Conference on Artificial Intelligence, AAAI (1982).

18.   Shaffert C, Cooper T, Bullis B, Kilian M and Wilpot C: 'An introduction to Trellis/Owl', Proc Conference on Object Oriented Programming Systems, Languages and Application, ACM (1986).

19.   Touretzky D: 'The mathematics of inheritance systems', Morgan Kaufmann (1986).

20.   Stroustrup B: 'The C+ + programming language', (2nd edition), Addison-Wesley (1992).

21.   Hamer J, Hosking J and Mugridge W: 'Static subclass constraints and dynamic class membership using classifiers', Tech Rept University of Auckland (1992).

22.   Augeraud M and Freeman-Benson B: 'Dynamic objects', Proc ACM (1991).

23. Hamer J: 'Un-mixing inheritance with classifiers', Proc Workshop on Multiple Inheritance, ECOOP (1992).

24. Frontini M, Griffin J and Towers S: 'A knowledge-based system for fault localisation in wide area networks', in Krishnam I, Zimmer W (Eds): 'Integrated network management', North Holland (1991).

25. IntelliCorp: 'The knowledge engineering environment', Menlo Park, California (1984).

26. Brownston L, Farell R, Kant E and Martin N: 'Programming expert systems in OPS5', Addison-Wesley (1985).

27. Patel-Schneider P, Brachman R and Levesque H: 'ARGON: knowledge representation meets information retrieval', Fairchild Technical Report 654 (1984).

28. Doyle J and Patil R: 'Language restrictions, taxonomic classifications and the utility of representation services', Artificial Intelligence (1991).

29. Litman D and Devanbu P: 'Clasp: a plan and scenario classification system', AT&T Bell Laboratories Report (1990).

30. McGregor R and Bates R: 'The loom knowledge representation language', Technical Report ISI/RS-87-188, USC/Information Sciences Institute (1987).

31. Nebel B: 'Reasoning and revision in hybrid representation systems', Lecture Notes in Artificial Intelligence 422, Springer Verlag (1990).

4

A MODEL-BASED APPROACH TO NETWORK, SERVICE AND CUSTOMER MANAGEMENT SYSTEMS

R Shomaly

4.1 BACKGROUND

Consider the impact of trying to develop large scale telecommunications and services management systems that are supported by powerful and complex computer systems. Powerful, in this context, encompasses, among other attributes, efficiency, usefulness, task simplification, integration and consistency. Complexity embraces distribution, multivendor, multiparadigm and evolving domain. The issue is how these systems are developed and maintained in an integrated and consistent fashion. To cope with the potential scale and complexity of these systems, a coherent and focused approach is required.

The ADVANCE project (part of the European collaborative RACE program) was set up to produce recommendations on how future large scale network, service and customer-based administration systems should be designed and developed. The project focused particularly on administration related issues, such as provisioning, planning and accounting. Prototyping solutions using advanced technologies have led to recommendations which have emerged in the project's final report [1]. This chapter is based on these results and recommendations. Other results of this project are recorded in Smith et al [2].

4.2   INTRODUCTION

This chapter describes several useful and usable concepts which have emerged as recommendations from the ADVANCE project. A number of factors influenced the emergence of these concepts. The objective was that these systems should be open, distributed, flexible, expandable, reusable and integrated. The results followed experimentation in each of the specification, design and implementation phases, and, therefore, the concepts proposed are practical experience.

The focus of this chapter is the concept of model-based management (MBM). MBM is a total management concept, encompassing behavioural representations of the management functions as well as information and data modelling at a high level. Modelling of managed objects has already been widely promoted by the concept of management information base (MIB) by the standards-making and similar bodies — ISO with OSI [3], ANSI in T1M1.4 [4], OSI Network Management Forum [5] and ETSI in NA4 [6]. The MIB is conceptually a repository of the data about the managed domain and mainly within the network level. The MIB in its narrowest definition contains managed objects which may support the definition of protocol data units (data structures for the definition of message sets which make up interoperable interfaces).

MBM has extended the scope of the MIB to include representations of the management information and behaviour in addition to representations of the managed system. This approach provides a total management view of the system which is both integrated and consistent. However, this definition has not been extended to include the role of people in management (for considerations on this area, see Dean [7]).

In this chapter a very broad interpretation of the term 'management information' is taken — it covers all activities associated with the management of a telecommunications network. These activities may include among others:

- pre-service activities including planning, data building, installing and commissioning;

- in-service activities including provisioning, maintenance, statistics gathering, traffic and performance management and billing;

- future service activities including performance analysis, forecasting and requirement identification.

A more comprehensive definition of management information and network/service management requirements has been given in Williamson and Azmoodeh [8].

The discussions in this chapter will cover both the approach to providing a consistent and integrated management system through the concept of MBM, and also the conceptual modelling of management and managed components through objects and providing transparent access to underlying systems resources (such as information sources, databases and network elements) as well as open access to other user applications, whether local or remote. Throughout this chapter the term 'user application' denotes management applications, customers, operators or any other applications which utilize the services provided by the management system.

The chapter also describes an object-based interface language that enables user applications to access the functionality of the conceptual model and the underlying management components. This language is considered to be a step forward, as it provides a comprehensive range of object query and access mechanisms using very simple constructs.

4.3   A TOTAL MANAGEMENT APPROACH

There are a number of issues to consider when developing a management system of a large and diverse nature. These issues include:

- the geographical distribution of the components, be they information servers or user applications;

- heterogeneity of entities due to multivendor sources and the diversity of development environments and platforms;

- providing the end users with a single, consistent and integrated interface to the underlying system resources such that the complexity of the system (i.e. distribution and heterogeneity) is abstracted;

- federation of independent, organizational systems — an important issue for multidomain networks is the heterogeneity of organizational policies (the impact of this issue on management across different domains has been addressed in the ANSA reference document [9]).

The problem of distribution leads to issues such as location, migration, access transparency, consistency, efficiency and security. The problem of heterogeneity leads to issues such as paradigm clash, consistency of representations, open computing and information access to multivendor systems and platforms. The issue of providing an integrated interface leads to issues such as expressive power of the language, the provision of an abstract model of the management system, the mapping of abstract objects to real entities, etc.

The major issue is how to provide an infrastructure in which the various components of the system are managed in an open, consistent manner and that the integrity of the system is assured. It is also necessary to abstract the complexity of the distribution and heterogeneity from the user applications and provide a common view of the system.

4.4   MODEL-BASED MANAGEMENT AS A SOLUTION

Model-based management is a term used to describe the concept of providing an abstract representation of the management and information system. This abstraction mechanism has been realized by an experimental development of a common information model (CIM). The CIM is therefore the main component of the realization of the MBM concept.

The CIM is conceptually a repository of all the management/managed information and functionality of the system. The information model is required to store, among other things, logical representations of network and system configuration, information related to customers and services, current and historic performance and trouble logs, security parameters and accounting information.

The model is a logical repository of information and is thus implementation-independent. Consistency and correctness of the model is ensured by mechanisms (consistency and integrity rules) so that only correct, consistent operations are applied to objects. Access to information and operations within the model is provided through a single object-based interaction language. This interface is discussed further in section 4.5.

CIM provides an intuitive conceptual representation of the real world through the close mapping between real-world entities and the logical objects. Its objective is to capture as many of the real-world semantics as are necessary to properly represent real-world entities and concepts (e.g. information and behaviour). The benefits of such a model are that:

- it provides a common view of the system to both application developers and end users, though the former will be more detailed;

- it provides a common means of access to shared information across diverse implementation platforms;

- it has the ability to represent all aspects of the system, including managed system, management information and management functionality in a single and coherent manner.

The managed system is the abstraction of resources which the system manages, e.g. customers, services, networks and network elements.

Management information in this chapter is defined as the information which the system requires to support management functions, e.g. call records, network performance records, service usage records and market trends and plans.

Management functionality is the behaviour of the system and details of invocation mechanisms and exported interfaces, e.g. functions which analyse network performance, analyse service trends and plan new service introduction.

It is possible, and in fact desirable, to take functionality out of the user applications and place it in the CIM. The advantages of this approach are that the model can ensure the consistency and integrity of the overall system through its well-defined semantics and is able to check that each operation is conformant with the semantics of the model.

The model provides a logical view of all user application interactions. The representations are logical and implementation-independent, thereby encouraging interoperability and open communications across management domains. Applications communicate using a single, common language and well-defined concepts. The CIM captures details of services, customers and networks each of which have their own schema within the overall framework of the model. This allows independent development of applications from services, customers and network details, thereby promoting reusability and automaticity.

The role of CIM within the system architecture is depicted in Fig. 4.1. Management applications view the underlying resources as logical objects with well-defined semantics and relationships. Management queries are posted (transparently through the distributed platform) to the logical objects as viewed within the CIM. The complexity of the mapping and retrieval of information is resolved by the CIM and the distributed platform.

4.4.1   Mapping of logical objects to real resources

CIM supports the user applications in performing their tasks. This support includes mechanisms for manipulating the real resources by manipulating the logical objects of the CIM. Consistency and correctness of each operation is ensured by consistency and integrity rules in the model, so that only correct, consistent operations are applied to the real resources. In addition alarms and other events, occurring in the real resources, update the model, which may in turn notify the user(s) of the change in circumstances. Thus, the model is kept up to date and mimics the behaviour of the system resources.

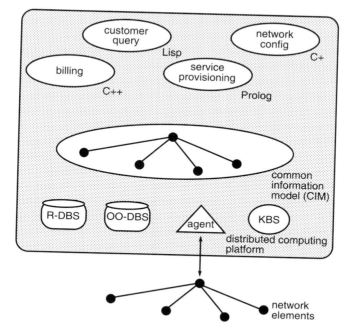

Fig. 4.1   CIM provides a logical view of the underlying resources (DBs, shared applications, network elements, etc) to the management applications. Distribution and access transparency are resolved by the distributed platform.

An object within the CIM is the view of (one or more) resource(s) within the management system. An object is defined by:

- its attributes;

- its relationships with other objects;

- the management operations that can be applied to it;

- the notifications emitted by it;

- the behaviour it exhibits in response to operations and notifications.

The object schema within the CIM provides an idealized view of the system to the users. Objects are used to provide an abstraction of one or many underlying informations and/or functions. Take, for example, a schema representation of a customer and its contracted services which may be logically represented as shown in Fig. 4.2.

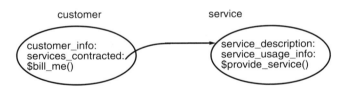

Fig. 4.2   Schema of a customer and their contracted services.

Such representations hide many unpleasant details of the underlying system set-up from the users. The mappings of logical properties to real resources may take the following form:

● customer__info is mapped on to one or more databases (relational tables) residing locally or remote (in fact, all resources are viewed locally through the distribution transparency service provided by the computing platform);

● services__contracted is a logical relation to the service object which in turn maps to many underlying components;

● $bill__me is a logical operation on the customer object which may involve the invocation of many applications and information entities both within and outside of the CIM.

The actual invocation and access to real resources is done through various application-specific handlers which provide the translation and access routines to these resources.

4.4.2   Modelling behaviour through objects

An object is a collection of data plus the operations used to manipulate the data. Operations invoked within an object may interrogate or change data within the object or invoke operations on other objects. This ability to produce changes of states within objects is named behaviour. Behaviour refers to state transitions and inducing change in dynamic properties of objects.

A second form of behaviour is changes in the state of the model due to external operations requested by the user applications. This behaviour exists when compound operations are provided by the object classes. The advantages of utilizing this type of behaviour is that it can effectively take functionality out of the user applications and place it within the model.

A third form of behaviour exists whereby the model has built-in rules that ensure that the changes made by the first two forms of behaviour are consistent with rules held within the model. Thus the model provides integrity and consistency rules to effectively keep the state of the conceptual model in line with the state of the managed resources.

4.4.3 Defining the model through the MODEL specification language

The importance of object oriented models in the design life cycle has been recognized by a number of standards and related bodies such as ANSI-T1M1, OSI, CCITT. These bodies recommend the use of models in the definition of data organization in the management information base (MIB), and in the definition of message sets and interface protocols.

In order to support the specification of an entity such as the CIM, or indeed to test its practicality, some means of defining and realizing the model was needed. For this purpose a modelling language was defined. It effectively enables the definition of the CIM and its realization in an object oriented deductive form.

For the implementation of the experimental CIM within the ADVANCE project, an object oriented logic specification language (MODEL) [10, 11] was developed. It was felt that the marriage of logic constructs with object oriented concepts provided a formal method for describing objects and their relationships that is precise and unambiguous. It provided well-defined semantics allowing concepts to be presented declaratively and transparently, i.e. all information is stated explicitly and is available for inspection. Consistency is an important criterion not easily met by an imperative object oriented model. Logic ensures consistency is maintained through the use of derived relations (inference) which avoid duplication of information and constraints that represent dependencies between values.

The specification language is capable of supporting a number of views of the information in the model. For example the composition of a network can be viewed from at least two different perspectives, a functional view and a physical view. The model attempts to capture physical, logical, functional and geographical notations by defining representative object classes at the top level of the specialization graph. Once such classes are defined it is possible to use the type concept associated with class in order to give good conceptual semantics to relations. For example the has_part relation represents the physical composition.

MODEL is an experimental language and has been created to facilitate the realization of an experimental network and service model within the CIM. A detailed account of the MODEL language can be found in Manning and Spencer [10] and Manning and Sparks [11]. The reader is also referred to

Chapter 8 for an object oriented approach of the formal specification language Z.

4.5   SYSTEM INTERACTION

Large and complex management systems such as network and customer administration systems have significant system interaction requirements associated with the collection, updating and management of large amounts of data by a multitude of co-operating applications and services. Databases, knowledge bases and object bases represent the information repositories with which such systems will most likely have to deal.

A pragmatic issue in the use of objects is the ability of user applications to access object structures. In areas such as customer, service and network administrations, there is a strong need for complex retrieval capabilities for structural information of management/managed objects.

The functional requirements of user applications coupled with other pragmatic issues such as operational requirements (speed, communications capacities, etc) make it necessary to move away from an object-at-a-time mode of access to a more powerful language where properties of many objects, as well as invocation of many operations, can be achieved through a single command.

4.5.1   Existing candidates

It could be argued that SQL (standard query language) could be used as the basis for such a query language. This necessitates that the CIM be relational. It is now recognized that relational models cannot handle the complex data structures needed for a complex management system and that is why the CIM is structured in an object oriented form.

Indeed, there are already a number of object oriented query languages (SQL-like, RELOOP [12], SQL ONTOS [13], etc). Languages which are extensions to SQL, seemingly have the advantage that they may be acceptable to users already familiar with SQL. However, as the relational and object oriented philosophies are fundamentally different, this advantage is only a superficial one. These languages lack the conciseness and simplicity which is required.

The suitability of CMIS/P as a 'higher level' communication medium is also doubtful. ISO system management CMIS/P [14] is defined to provide an interoperable interface among conformant management entities. CMIS/P is based upon an object/attribute/relation model of managed objects, it supports strict typing, it is standardized to be vendor-independent and its syntax is defined using formal notation of ASN.1. These together with the notions of manager and agent roles make CMIS a suitable interface between the network element and the operation system communications protocols. However, to support the demanding communications requirements of complex management systems, higher level communications services are required. An important limitation of CMIS, is the very limited power of its retrieval mechanisms.

The model of query representation in CMIS/P is based on the hierarchical structure of the naming tree which represents the containment relationships among network management objects. The more demanding communications needs of model-based management require a much more powerful language for interactions which involves arbitrary graph structures. The expressive power of CMIS is therefore limited with respect to models of MBM with complex interrelationships.

CMIS/P is not a full object query language. Access and manipulation of information is limited to managed object attributes and values, furthermore CMIS/P does not provide specific transaction processing facilities other than those which are provided by synchronization parameters.

In conclusion, although CMIS is based on the object oriented paradigm, it needs to be augmented to have a powerful query language capability to overcome the above-mentioned limitations and is therefore unsuitable as the interoperable interface language at the information and computational levels.

Formal languages as a candidate have also been inappropriate. Of the languages defined for interface and query languages the unifying basis of most well-founded approaches has been the first-order predicate calculus (FOPC). Although FOPC is well-founded with provably known power and limitations, it lacks the organizational and abstraction mechanisms which are helpful in reducing the complexity of the query language. What is needed is a data model and a language where individual items (objects, relations, attributes, etc) can be represented explicitly.

4.5.2   OBSIL

The CIM is logically centralized within the management domain. It acts as a source and sink of information and functionality to the user applications. Communications between the CIM and the user application is provided through a single, well-defined and well-structured object-based system-interaction language (OBSIL). All user applications and resources access the CIM using OBSIL. OBSIL was designed using object oriented methodology to provide a powerful, yet simple to formulate, interface to the CIM.

OBSIL is simple, concise and powerful. It provides mechanisms for expressing complex qualifications on objects. It is conceptually concise in the sense that it does not need artificial variable names. The query is formulated using path expressions to traverse the relationships of an object oriented model (implicit 'joins'). While traversing relationships, quantifications such as 'for all', 'there exists', 'at most n', etc, can be specified.

The OBSIL language is navigational in nature. The language provides mechanisms to access objects by navigating through the model and qualifying/quantifying attributes and/or relations of the model.

For example the request:

Retrieve customers who are older than 55 and subscribe to the 'tele-bingo' service

will be simply formulated as:

```
Get Customer.name
    such-that
        (age > 55
        and
        has__service→Service.name = 'Tele__bingo')
```

The language uses no artificial variables or explicit joins. It is argued that explicit joins are not needed as all the meaningful relations are represented in the object oriented model. The effect of this is that queries are much easier to read as they are much closer to natural statements.

4.5.3   Operations of OBSIL

OBSIL operations are **create, get, action, delete** and **modify**.

Create: the create operation is used to create new object instances. The create operation returns the identifier of the object created or an error indication if the operation fails. For example:

```
Create Service
    {attributes name = 'Video Telephony Service'
               quality = 'high'
    relations is-provided-to = [CUST001, CUST002, CUST003]}
```

Get: the get operation is used to access object instance information (for details, see examples above).

Action: the action operation is used to invoke operations (methods) on objects. The return result of an action is dependent on the defined result type of the object method. An error indication is returned if the operation fails. For instance, the following command:

> **Action** Service.$provide-service (CUST001)
>   **such-that**
>   quality = 'high'

invokes the method $provide-service( ) of all high-quality service objects for the customer CUST001.

Delete: the delete is used to remove object instances. The result of a delete operation is a list of deleted object IDs or an error indication if the operation fails. For instance the query:

> **Delete** Customer **such-that**
>   lives__at→Address.town = 'London'

deletes all objects that represent customers who live in London.

Modify: the modify operation is used to alter information within existing objects. The result of a modify operation is the instance IDs of the altered objects or an error indication if the operation fails.

For example, the following query sets the address of the subscriber CUST001:

> **Modify** Customer.lives-atAddress→
>   [town = 'Cambridge'
>   street = 'Oak rd'
>   no = 13]
>   **such-that**
>   id = CUST001

For the full detailed account of OBSIL the reader is referred to Azmoodeh and Shomaly [15, 16] and Shomaly [17].

4.6   CONCLUSIONS

Adopting the model-based management approach has proved useful in the experimental development of network, service and customer administration systems. This approach was found to be not just useful, but necessary in the development of the management system. Although a great deal more work is required in order to fully prove that MBM can provide a sound basis for the design and development of complex management systems, it is felt that the prototyping work within the ADVANCE project has provided a very satisfactory basis for advocating that MBM can be used as a sound framework for building such systems.

It is believed that the MBM concepts provide a very powerful means of understanding and structuring the problem of building large and complex systems, because it allows the builder of such systems to realize the problem through a model which represents as closely as possible information and knowledge on the actual physical telecommunications domain.

The central component of the architecture which supports MBM is the common information model (CIM). This can be seen as a single logical entity that contains the shared management information and functionality of the system. Therefore, for the application developer, complexity will be reduced because considerable domain knowledge can be incorporated into this central information base rather than in individual applications' private information bases. Because this information base is close to the human conceptualization of the domain, it simplifies the application developers' understanding of what resources to use and how to use them.

The CIM supports the information requirements of a number of heterogeneous applications by providing abstract views of shared information and ensuring mutual consistency of these views. It facilitates integration and reusability across the management system.

The MODEL specification language has proved to be capable of representing the structural and behavioural semantics of the objects within a management domain. It was used to specify an experimental network and service model within the CIM.

OBSIL is a high-level object manipulation language which is designed to support the interaction requirements across the management components. It was used to provide access to the CIM as well as facilitating generic access to multitechnology database systems (Oracle, Informix and ONTOS) and also for providing access to network elements (through translation into CMIS).

The importance of providing a consistent and integrated approach to telecommunications management has been recognized by BT for some time. This emphasis is reflected by many new projects which aim to provide a single and consistent interface to the underlying BT services which are in many cases very large and complex. Currently, as an example, a single query from a customer may need to be manually propagated on to many information bases using separate platforms.

The concepts described in this chapter have been used and evaluated within the RACE-ADVANCE project of which BT has been a collaborative partner. These concepts are considered to be an important step towards integrated management of the telecommunications networks and services which will be emerging in the very near future.

REFERENCES

1. Race Project ADVANCE (R1009) Final Report, ADVANCE Deliverable No 4 (December 1992).

2. Smith R, Mamdani E H and Callahan J (Eds): 'The management of telecommunications networks', Ellis Horwood (1992).

3. ISO International Standards Organisations: 'Informational retrieval transfer and management for OSI — system management overview', ISO/IEC DIS10040 (1990).

4. ANSI: 'Modelling guidelines', ANSI T1M − .5/88-014R4 (December 1988).

5. Embry J, Manson P and Milham D: 'An open network management architecture: OSI-NM forum architecture and concepts', IEEE Network (July 1990).

6. ETSI NA4, Generic Network Model (1990).

7. Dean G, Hutchinson D, Rodden T and Sommerville I: 'Distributed systems management as a group activity', IEEE 1st Int Workshop on Systems Management, Los Angeles (April 1993).

8. Williamson G W and Azmoodeh M: 'The applications of information modelling in telecommunications management networks', BT Technol J, $\underline{9}$, No 3, pp 18-26 (July 1991).

9. ANSA APM/RC366.00, ISA project: 'Federated Platforms' (July 1992).

10. Manning K and Spencer D: 'Model-based network management', 4th RACE TMN Conference, Dublin (1990).

11. Manning K and Sparks E: 'Service and network model implementation', 5th RACE TMN Conference (1991).

12. Cluet S et al: 'RELOOP, an algebra based query language for an object oriented database system', Data and Knowledge Engineering, $\underline{5}$, pp 333-352 (1990).

13. ONTOS DMBS, Ontologic (1992).

14. ISO/IEC DIS 9595—2, and ISO/IEC DIS 9596-2, ISO-OSI CMIS/P.

15. Azmoodeh M and Shomaly R: 'An object based system interaction language as a basis for TMN systems interactions', 4th RACE TMN Conference, Dublin (1990).

16. Azmoodeh M and Shomaly R: 'OBSIL: an object based system interaction language', RACE-TMN Object Oriented Modelling SIG proceedings (June 1991).

17. Shomaly R: 'OBSIL user guide version 4.0', RACE project ADVANCE 1009 Deliverable 5 (October 1992).

5

OBJECT ORIENTATION — EVOLVING CURRENT COMMERCIAL DATA PROCESSING PRACTICE

A G C Heritage and P G Coley

5.1   INTRODUCTION

Over the last few years, object orientation (OO) has become of increasing interest to commercial data processing practitioners. One reason for this is that the traditional profile of commercial data processing is changing. Increasingly systems are being accessed by other systems, and distributed processing has become an issue. Another reason is that there is demand to reuse existing functionality, and a business need to change parts of the business functionality quicker than present methods allow. Object orientation has the reputation that it can help meet these new needs in a way, it can be argued, that traditional methodologies cannot match. This reputation led BT to explore exactly what object orientation might offer for commercial data processing systems.

This chapter describes the issues that faced the authors in adapting published work in object orientation to support the development of large systems. Such systems typically use COBOL with network and relational databases as their implementation vehicle. Therefore the chapter concentrates on three issues:

- the adaptation of the COBOL environment;

- dealing with legacy systems[1];

- how, in our view, the basic object orientation approach has to be enhanced to support the construction of large systems.

5.2   ENVIRONMENT DESCRIPTION

It is helpful to understand the environment into which object orientation had to be introduced. First, there was a strong tradition of analysis and design, and systematic methods, such as SSADM and information engineering, had been adopted over the last ten years. These methods are datacentric, and any systematic investigation of a new system would always begin with an analysis of data. Heavy investment had gone into rationalizing data structures, and creating specialist roles such as data and database administrators for their maintenance. Furthermore, even at implementation time, data continued to dominate — the performance of a system was usually controlled by the speed of database access.

Linguistically, whereas there were moves to program in object oriented languages, particularly for the client portion of client-server applications, the fact was that a large amount of programming was and would remain in traditional imperative languages. Chief among these was COBOL. To COBOL could be added similar proprietary languages which were components of code generating CASE tools. Ways would have to be found to implement object oriented constructs in these languages. If such solutions were not available, and the received wisdom was that they were not, then the future of object orientation in commercial data processing would be bleak.

Neither was there a free hand in the choice of database. Whatever the scope in the future for object oriented databases, the fact is that today the only type of database accepted for high volume work in BT is the network database, with relational databases being used where volume is lower and response less crucial. Therefore, ways would have to be found to translate

---

[1] Legacy systems are core systems which have been constructed over the last 10 years or so, and upon which the business will be dependent for some time yet.

object oriented designs into constructs that these types of database could handle.

Finally, the management style of the organization was towards pragmatism. There was a desire to use and adapt our present practice rather than undertake wholesale replacement of the present investment by something new.

5.3   ADAPTATION OF THE COBOL ENVIRONMENT

In talking about COBOL, we were concerned with IBM's implementation of it because that is what, in general, BT uses. This implementation is known colloquially as COBOL II, and we used, for reference, IBM's publication 'VS COBOL II Application Programming: Language Reference' [1]. VS COBOL II is a subset of the COBOL 85 ANSI standard, but has some extensions. From here on, references to COBOL in this chapter refer to this IBM implementation.

COBOL has no explicit support for object orientation. Therefore it was a question of seeing what structures could be built into the language to support object orientation. The issues were internal object structure, inheritance, run-time instantiation, storage persistence and message passing. The chapters in Rumbaugh et al [2] on working with non-object oriented languages provided us with a starting point for this work.

5.3.1   Internal object structure

In creating an internal structure for an object, there has to be some way to describe its data structure, its method selection mechanism and the method structure itself. Because of its history, COBOL has excellent ways to represent data structures. However, all data definitions have to be made in the 'Data Division', only one being allowed per program. So, if a program contains more than one object, the data definitions cannot be co-located with the objects' processing code. While this is a pity, documentation could alleviate the problem, and we did not regard it as a serious drawback.

Turning to the method selection mechanism, COBOL has plenty of devices such as 'section' and 'paragraph' for partitioning programs into objects and their methods. The CASE programming structure is used typically for method selection within an object. COBOL supports this structure ('evaluate' in COBOL), so there is no problem here. COBOL also has plenty of options for run-time organisation. It allows any shade from the compilation of relevant methods into a single run-time module, to the compilation of methods into individual run-time modules. A particularly flexible feature of COBOL in this respect is the ability to declare multiple entry points (the rarely used 'entry' verb) in a called module, thus avoiding the CASE programming statement. So, for method selection, COBOL offers every support required for an object oriented approach.

For the method structure itself, we devised a code template (Table 5.1).

Table 5.1   Method structure code template.

1	Reference data (method name; logical description; volume)
2	Interface definitions
3	Pre-conditions
4	Method body
5	Post-conditions

In itself the template was unremarkable, perhaps even trivial. But its very obviousness made it powerful — practitioners could intuitively see from it how a method was constructed, how methods fitted into the object oriented paradigm, and where the documentation should be placed. COBOL offers no facilities to enforce the proper use of such a template, but on the other hand, as there were no difficulties in implementing this template in COBOL, the language seemed in our view an adequate vehicle with which to program methods. In summary, therefore, COBOL seemed to be able to give good support for a proper internal object structure.

5.3.2   Inheritance

We found no way to properly implement inheritance. The only possibility seemed to be to break out the code aspects of the class structure into separate modules. Messaging traffic would always be directed at the level of most specialization. Therefore, referencing to greater levels of generalization would be a call up the hierarchy.

5.3.3   Run-time instantiation

COBOL has no way of supporting instantiation of objects at run time. It is limited to creating a fixed number of images of the object type at run time, and persistent instances have to be imported from elsewhere to fill those images. A typical device would be to use arrays, so that the actual class method code could operate on any instance by referring to the appropriate part of the array where the instance lived.

5.3.4   Storage persistence

The popular way to provide persistent storage for COBOL is to use a database. Representing the database record structure is no problem as there is an explicit match between the COBOL definitions in the 'file division' and the particular database management system structures. Also, the interpretation of objects into database designs is straightforward — we used the work of Rumbaugh et al [2] for relational databases and adapted it to cover network databases. However, an object's attributes may have to be represented at implementation time by a mixture of persistent data from a database and data that is only required during the period of manipulation held in 'working storage'. Moreover, the object oriented approach of being able to alter, say, single attributes at a particular time can only be provided by hauling the whole of the database record into store, as database records can only be read or updated in totality. Therefore a strategy had to be found to associate the working storage with the database record pulled into store to form the complete set of object attributes. The strategy had two parts. The first was the notion of 'activating' the object. This required a pre-condition to method execution. It consisted of first allocating an area of non-persistent store, then initializing the area to ensure it was free of data from previous use, and finally retrieving the persistent record from the database. The second part dealt with the problem that when the database image has been brought into store two copies exist, initially in the same state. 'Passivation' of the object controlled this. It consisted of marking an object as ready for persistent storage update, and ensuring that, when that had happened, its allocated non-persistent storage area would be freed.

5.3.5   Message passing

The nature of COBOL makes the passing of parameters between objects difficult. While between modules a rigid parameter passing scheme can be exploited, within a code module no such facilities exist and all variables are global. Even so, it is possible to support rigorous encapsulation providing one is prepared to develop the software. Some possible ways forward are discussed here.

One solution within a module is for the programmer to ensure that although variables are defined globally, they are only used in a local manner. Following this approach, we found it possible to implement full encapsulation by using 'message' variables. To send a message, values would be placed in these variables, and the called routine could only extract the data from them and them alone. The problem is that the system could be hopelessly abused because there was no way of truly implementing data hiding — a called object's attributes could always be accessed by the unscrupulous programmer. To deal with the occasions where it was better to break encapsulation, the idea of 'friend' (where two objects are related such that one object may access the attributes of another) was borrowed from C++. Implementing a 'friend' amounted to a design issue where the developer recorded where it was permissible to break encapsulation.

To overcome the rigid parameter passing scheme between modules a tagged parameter scheme was developed. At its simplest, tagged data is passed by a method placing it in a known area of memory, which can then be read by another module. Tagging allows arguments to be in any order, and for different sets of parameters to be supported by the same method selector. The value of this is that different versions of the recipient object can be supported, at the cost of using a horrendously complex piece of code (several hundred lines to complete a call).

5.3.6   COBOL — summary

Our conclusion about COBOL is that it can support any object oriented feature to some degree, with a cost ranging from the trivial to the severe. For instance, creation of a structure to select a method is easy, as is the creation and calling of the method code. At the other end of the scale, the lack of in-built parameter-passing mechanisms makes a watertight run-time encapsulation system something of considerable complexity. Therefore, in practice, implementation in COBOL can really only reach a level which could be described as object based, not fully object oriented.

5.4   DEALING WITH LEGACY SYSTEMS

There are rarely greenfield sites available for the implementation of a new paradigm. Indeed, many 'new' systems now tend to consist of the new plus the old bundled together. Therefore we had to consider how existing legacy systems could be incorporated into an object oriented universe. There was the well-known technique of placing a wrapper around a legacy system to provide the interfacing and data-hiding to make it appear object oriented. However, this was not always sufficient. There were cases where there was a desire to extract some of the functionality from a legacy system, and repackage it as an object for use by both the donor system and other systems. Scanning of the literature had revealed little information on how this process of decomposition and recomposition might be done. Thus we had to start from scratch. Below we outline the rules we developed to achieve this extraction. The process is divided into analysis and design.

5.4.1 Analysis

5.4.1.1 Identifying candidate objects from data structures

COBOL has the immense advantage that it has a data division in which the major data structures are laid out. So the code could be scanned to reveal which of those data structures were used, and they became candidate objects. These objects were then surveyed to create more 'logical' data groupings based upon business function. A particular aim here was to associate all working storage variables with more persistent (i.e. database) records. The result was an object data architecture which could consist of several traditional data entities plus some working storage. Skeleton 'sections' (a COBOL code partitioning device) were set up for each of these objects. These 'sections' would be populated with code in later stages of the process. We did not encounter any scope for inheritance. This may be a consequence of reverse engineering COBOL code or it may be that the sample analysed was not congenial in this respect — we do not know.

5.4.1.2 Dealing with code

The process of allocating the functional code to the data oriented objects was one of gradual refinement. The start point was to examine each 'section' of code, and judge to what data object its function suggested it belonged. That judgement made, the code would be physically moved to the owning object's 'section' (discussed above), and transfer of control statements put in to retain the program logic flow.

Then the control structures within the code were examined. We believed that the assignment code that lay between conditional statements was a chunk of function that probably constituted a candidate method. Deciding what object it belonged to was assessed from a mixture of considerations regarding the variables tested before a piece of code was executed and what variables were assigned to in the assignment statements. If there was a strong leaning towards one object evidenced in both the controlling statement and in the assignment statements, then the code would be allocated to that object. Rules were also designed for weaker cases, that applying to the weakest case being sheer judgement! The re-allocated pieces of code were physically moved in the way discussed earlier.

The result of analysis was to achieve a set of objects with co-located candidate methods, some of which were duplicates or very similar. The transfers of control in the program were likely to become messages, but the parameters were unknown. They were hinted at by the breaches of encapsulation, which at this stage were endemic.

5.4.2   Design

5.4.2.1   Deciding the degree of encapsulation

One of the targets of design is to bring about encapsulation, the options for which were described earlier. To help in choosing, we devised rules for selecting which option to take. For example:

- if an attribute retains its value from call to call, it is part of the state of the object, and access is encapsulated;

- if an attribute is set on each call to the object, essentially it is a messaging variable, and it is suitable for use as a 'friend'.

5.4.2.2   Removing duplicate and similar code

Likely duplicate code was spotted by manual inspection, given that the methods belonging to an object were now co-located. Duplicate code was replaced by replacement calls to the surviving code. Similar code was more difficult. As code has grown over the years, one pattern of enhancement is to take the original code and add in extra statements. Thus some original COBOL statements like:

```
Move X.1 to A.1
Move X.2 to A.2
Move Y.1 to A.3 etc ...
```

would typically be targeted at one object, A. Later the code might have become replicated and enhanced to become:

```
Move X.1 to A.1
Move X.1 to B.1
Move X.2 to A.2
Move X.2 to B.2
Move Y.1 to A.3 etc ...
```

Where the ordering of these statements was not important, and this was usually the case, our solution was to de-interleave the statements targeted at *A* from those targeted at *B*. The code targeted at *B* would be moved into *B*, and the code remaining in *A* would then probably be removed as a duplicate.

5.4.3   The result

The original program had become a hierarchy of calls, with most of the functionality in the bottom layer. Therefore we were able to explain in ten minutes how the program operated — that would not have been possible with the original code where the concerns of the program could not be clearly seen. We could see the objects clearly, and we could see the nature of the traffic between them. Moreover, now that they had a clear interface, they could be reused if required. Table 5.2 shows the statistics.

Table 5.2   Program code statistics.

Size of original code, in lines	235
Original lines removed as duplicate or surplus	61 (26%)
Lines of original code retained	157 (67% of original)
Lines amended	17 (7%)
Lines added to implement encapsulation and better control structure	138
Size of new code, in lines	312 (33% increase over original)
Time taken	40 mhr per 1000 lines (reduced by 75% with automation)

The experiment showed that objects could be extracted for use elsewhere. Although the total code size expanded, this was not too alarming because it was due to implementing encapsulation. We would expect the initial overhead to reduce as exploitation of reuse increased. However, inspection of the code revealed that object referencing went down as many as seven layers.

While this manifestation does not conflict with object oriented theory, from a practical point of view of testing and maintenance it was felt to be a problem. Another area of doubt was the size of methods, which averaged eight lines. Scaling up to a not untypical size of system, this would lead to about 5000 methods. This seemed too many to handle comfortably.

5.5   LARGE PROJECT ISSUES

The potential for object orientation for large systems was seen as:

- promoting the reuse of code;

- avoiding the development of monolithic code;

- allowing functionality to be added and replaced with minimum effort.

However, it was by no means automatic that these desires could be achieved for large systems. We were not able to find substantial pieces of work that addressed the issues of large numbers of objects involved in large systems. We felt that this was going to be a significant problem, and would be for a number of reasons. For instance, we already noted in the work on reverse engineering that referencing could go down to any number of levels. However theoretically acceptable that might seem, practitioners had strong objections on testing and maintenance grounds and we had to provide a more acceptable transaction structure. Other issues were that the paradigm was able to incorporate requirements such as screen and report design (important to most data-processing systems), that it should allow systematic analysis of the logical models from a performance point of view, and that there should be a way of checking that business functions were properly implemented. Traditional methods such as SSADM provided these features, and we had to ensure that any object oriented approach offered equivalents. We discuss below some of the answers at which we arrived.

5.5.1   Object representation of the human interface

Screens, reports and menus consist not just of the image, but also of the structure by which they are invoked, including the consequences of invocation, usually a business event of some sort. Our solution was to consider a group of menus plus their navigation structure as one object, with the different menus displayed being expressions of different states. Movement from menu to menu was accomplished by internal methods calling each other, and when selection was completed, events/messages would be triggered to drive other objects. This is an approach which works for simple selection structures. We are not convinced that it would be the right way to model window environments, because, as each window could have many states, it may be better to model each of them as a separate object, with the navigation structure, essentially the message structure, binding them into a complex object. For a full discussion regarding choice of appropriate object granularity, see Chapter 6.

5.5.2 Transaction structure

A commercial data processing 'transaction' may be viewed as the sequence of operations required to carry out a business function usually triggered by a business event of some sort. Traditionally transactions have bundled together both the sequence of control and the considerable amount of processing necessary to deal with the raw data. As objects have their own internal state and processing, there is less need for this to be part of the transaction — the object would hold most of the system complexity, such as tuned SQL for database performance. This drove us to consider that this should be taken to the limit and a transaction should be confined to control, with other objects invoked to provide it with services. This view echoed that of Coleman [3]. He discussed the difficulty of tracing consequences through a large object model, and suggested the idea of a controlling object as an organizing convention. This confirmation of the idea allowed us to consider a data processing transaction to be like a controlling object. This model is illustrated in Fig. 5.1.

Four passive objects are shown being used in Fig. 5.1. The first collects a key value (CustID) from a screen. The transaction hands this value to the

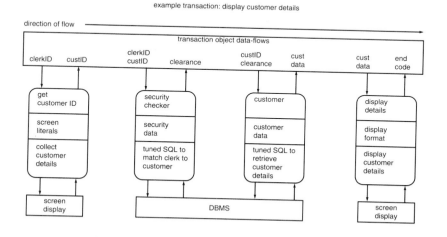

Fig. 5.1   Transaction model.

second object, which then derives a value (clearance) for the transaction to interpret. The third object uses two values the transaction has collected to drive a database interrogation. The results are returned to the transaction which passes them to the fourth object for display.

The difference between this and the canonical expression of an object oriented design is as follows. In the canonical model, the first object might of itself have consulted the second to collect the clearance. Also, the third object may itself have been able to drive the fourth object to display the data it retrieved. These approaches are neater in that data handling is reduced. The disadvantage is that this way of retrieving data by using intermediate objects results in a more obscure design. In our flatter structure, the sequence of events necessary to carry out the transaction is very visible, and the passive objects are simple and easily reusable.

5.5.3   Enriching the models

A survey of the literature showed many techniques and models for object oriented analysis and design. In our view, they distilled into three implied perspectives covering data, dynamics and function. However, when closely analysed the techniques could generally be classified into four types of model. These were class relationship (representing data), object behaviour (representing the internal dynamics of an object), method internals (representing the function), and object communication (representing the external dynamics or how objects work together). In Table 5.3, we analyse how five published techniques support the four model types.

Table 5.3   Object oriented analysis and design models.

	Class relationship	Object behaviour	Method internals	Object communication
Booch [4] [5]	Class diagram	State transition diagram	Operation template	Object diagram Timing diagram
Coad and Yourdon [6] [7]	Class and object diagram Subject areas	Object state diagram	Services	Message connections
Rumbaugh et al [2]	Object model	Dynamic model	Functional model	Event trace Event flow diagram
Shlaer and Mellor [8] [9]	Information model	State model	Process description	Object communication model Action data-flow diagram
Wirfs-Brock et al [10]	Class and responsibilities	No model	Responsibilities and contracts	Collaborations and protocols

Most of the authors cited in Table 5.3 provide techniques for each of the four model types. However, none of them make explicit the fundamental importance of the split between the internal and external dynamics of an object. To those involved in commercial data processing, where internal states of individual objects are often complex, this distinction is important and has to be explicit. Also the four-model approach gave homes for the extra analysis and design information which needed to be included. Below we discuss each of the model types, and how we have exploited them.

- Class relationship (model 1)

  This model consists of the object class, its attributes, the static, data oriented relationships between objects, and the names of their methods. The model is common to most of the published literature in some form and we use it in the same manner. The only addition is that, as a contribution to analysing performance, we would consider this model to be the correct place to record information about the amount of traffic the object handles.

- Object behaviour (model 2)

  This model records how the object changes through time, and what are the permissible paths from one state to another. It is events, either internal to the object, or received from outside, that cause the object to move from state to state, and they would be recorded here. Again, most of the published methods support this model. However, there seems to be some debate as to its usefulness. For instance, Coleman et al [11] suggests that this model is not essential. Our experience to date is that in commercial data processing it is important in discovering the type of problem like a telephone bill for £0!

  In addition, we saw two other uses for this model. One was to secure reuse. There had to be some sort of vehicle around which browsing for reuse could be hung, and the behaviour model provides a simple library of the methods which affect an object. The second use was to help ensure that all the required methods have been identified. The behaviour model contributes to securing this completeness because it provides a logical framework for relating methods to each other and allows missing ones to be noticed.

- Method internals (model 3)

  The standard view of method internals is to show how a method works. It is the inward-looking perspective, and it records the things that the object is capable of doing, how they are done, and the interfaces to the outside world.

- Object communication (model 4)

Most objects in themselves carry out a fragment of a task — many objects collaborate to complete a particular task, or as we have termed it, a transaction. One purpose we see for the object communication model is as a trace for each transaction to verify that it is carrying out the business function that it is supposed to do. In addition, carrying out such a trace ensures that at every point in the transaction where a database access is required, there is sufficient information known, such as a key or other adequate parameter for the access to be executed. Gaining these types of assurance is critical in commercial data processing system design, both to customers and technicians.

The object communication model also assists with issues of database performance, which are well covered by traditional structured methods, but not by object oriented methods. For instance, if an object controls access to a database, what matters is the aggregate demand on that object to supply information which then has to be analysed so that appropriate design decisions for performance can be taken. We see traversal of the object communication model providing the access paths and the data requirements of individual transactions. The aggregated effect of these transactions would be held in the class relationship model, as mentioned earlier. However, it is the two perspectives together that give the complete logical picture of how a system is likely to perform.

It can be seen from the above discussion that the models interact, events being the vehicle of that interaction. In looking at a real-world transaction, first of all a business event will trigger a message to an object (model 4). The object will then check its state (model 2), and, if required, carry out a method (model 3) which acts upon data (model 1). In doing this the object changes state (model 2), and may also choose to use the services of another object (model 4). This choice triggers activity in the other object (model 4) and the pattern repeats itself in that object.

The attraction of this set of models is the decomposition of the problem that it offers. For instance, while traditional structured methodologies deal with the interactions of state, our experience is that they use annotation systems which are too complex for the average practitioner to use. On the one hand, this approach allows the problem to be seen in the small (analyse the object through time), and the autonomy of the object (I refuse to do this because I am in the wrong state) allows infringements to be dealt with purely within the bounds of the object. On the other hand, the business event and its consequences allow the model to be examined in the large. We have no views as to whether the models should be developed in any particular order.

5.6 CONCLUSIONS

We set ourselves the task of seeing whether object orientation could be used to support a commercial data processing project. The answer to that is 'yes', but it is not necessarily easy.

We were able to show that COBOL can be made to support enough of the concepts of object orientation to make it worth using.

We have been able to show that reverse engineering from a traditional to an object oriented design is possible. This eases the problem of what to do with legacy systems and enables object orientation techniques to be used.

We do not think that the commercially available OO methodologies are adequate in themselves for the problems of developing large-scale commercial data processing systems. However, if they are viewed in terms of the four-model approach, and supplemented with extra information on issues like performance and transaction structure, we think that they can work. The four-model approach will be trialled shortly in order to verify this view.

REFERENCES

1. International Business Machines Corporation: 'VS COBOL II application programming: language reference', (GC26-4047-3) Fourth Edition (1986).

2. Rumbaugh J, Blaha M, Premerlani W, Eddy F and Lorensen W: 'Object oriented modelling and design', Englewood Cliffs, New Jersey: Prentice-Hall International (1991).

3. Coleman D: 'Object-oriented analysis model: what could be more precise?', an invited talk at SCOOP Europe (1991).

4. Booch G: 'Object oriented design with applications', Redwood City, California: Benjamin/Cummings (1991).

5. Booch G: 'The Booch method: notation', Rational (1992).

6. Coad P and Yourdon E: 'Object oriented analysis', 2nd Ed, Englewood Cliffs, New Jersey: Prentice-Hall International (1991).

7. Coad P and Yourdon E: 'Object-oriented design', Englewood Cliffs, New Jersey: Prentice-Hall International (1991).

8.   Shlaer S and Mellor S: 'Object oriented systems analysis', Englewood Cliffs, New Jersey: Prentice-Hall International (1988).

9.   Shlaer S and Mellor S: 'Object life cycles, modelling the world in states', Englewood Cliffs, New Jersey: Prentice-Hall International (1992).

10. Wirfs-Brock R, Wilkerson B and Wiener L: 'Designing object oriented software', Englewood Cliffs, New Jersey: Prentice-Hall International (1990).

11. Coleman D, Jeremaes P and Dollin C: 'Fusion: a systematic method of object oriented development', Hewlett Packard Laboratories (1992).

6

ENCAPSULATION — AN ISSUE FOR LEGACY SYSTEMS

E S Cordingley and H Dai

6.1   INTRODUCTION

BT currently uses a number of large, well-established database systems which are written in languages like COBOL. Their data structures and operations were designed to take full advantage of performance efficiencies of database technology current at the time of their development. These are valuable, and in some cases business-critical, assets, but they pose problems for the future.

Bloor [1] paints a gloomy picture of how such systems age. The business needs they were designed to satisfy are subject to change and additional business requirements arise. The necessary addition and modification of code make them larger and more difficult to maintain, their logic deteriorates and the elegance and coherence of their original designs are increasingly compromised. Nonetheless, until there is an economic case for their redesign, restructuring, or replacement, they will continue to be crucial to the business. As a result there is an urgent need to find ways of adapting these legacy systems.

Object technology — a term which avoids the need for distinguishing between object based and object oriented (OO) developments — is maturing and appears to be a promising source for industrial strength solutions of the kind that are needed [2-5]. An impetus for object technology came from noticing that the objects we use change rather less than the way we use them. This relative stability of objects is why — in systems which take full advantage of object technology — requirements, design, implementation, maintenance and enhancement are all based on objects. As the pervasiveness of objects means the design of such systems is less likely to be compromised than conventional systems, and their logic less likely to deteriorate during their operational life, they age more gracefully than non-object technology systems [1]. As discussed later, object technology provides possibilities for both additional flexibility in the short term and ways to migrate large legacy database systems to more future-proof forms over the long term.

It is not usually possible to take full advantage of object technology with BT's existing, legacy systems because the languages in which they are written do not support crucial features of object orientation such as inheritance and polymorphism (Chapter 1 gives a fuller explanation of these terms). Legacy systems cannot usually be metamorphosed into true object oriented systems without radical reprogramming. At present, this is regarded as prohibitively costly, so other strategies must be considered to get the benefits from object technology. Possibilities include putting object oriented/based wrappers around systems or modules, introducing object oriented front ends, creating object oriented modules which interface with them, or using just those features of object technology which can be introduced into the existing system without having to redesign it or rewrite it. Chapter 5 describes how notions of encapsulation can be applied to COBOL systems bringing cost-effective benefits without necessitating major redevelopment.

Although benefits from object technology are available even for systems written in non-OO languages like COBOL, they will only be realized if appropriate choices are made when designing and implementing the software. It is vital, therefore, that object technology notions are well understood and choices are clear. This chapter reports results from the project which explored the notion of encapsulation, a powerful feature of object technology which can be used in non-OO systems as a first step in gaining the benefits of the kind outlined in Chapter 1.

In section 6.2 the many sides of encapsulation are discussed and a framework provided for analysing the encapsulation strategy adopted in a system. In section 6.3 the framework is applied to an example from COSMOSS, a BT development. Section 6.4 presents conclusions from this work.

6.2   ENCAPSULATION CRITERIA

Berard's survey [6] of the terms 'abstraction', 'encapsulation' and 'information hiding' provides an important clarification of the notion of encapsulation. He identifies its essence as the process or the product resulting from 'the enclosing of one or more items within a physical or logical container' and notes that this implies nothing about the nature of the 'walls' of the enclosure. Fundamentally, encapsulation allows a number of items to be regarded as a single thing which can be reasoned about as a whole, but, strictly speaking, that is all. Anything else is an enrichment of the essential concept.

This explains much of the confusion about encapsulation. In common parlance, as in object technology literature, the notion of encapsulation is usually loaded with implicit meanings about the 'walls' of the enclosure, how much they hide and how well they protect what is enclosed. The exact implication of the term differs dramatically among users, as Berard's chapter shows. This is not usually a problem in general conversation. It becomes a problem, however, when the notion of encapsulation needs to have practical power, as it does in the context of software development. Choices about how encapsulations are defined and used may crucially affect a system's integrity and performance. To have technical utility the notion needs to be unpacked so that separate aspects of it can be treated explicitly and each can play its appropriate role.

Our project adopted Berard's definition of the essence of encapsulation and, like him, used it in both its senses, as process and as product. Since all one can be sure is being implied by the term is its essence, all the rest were treated as optional enrichments of the term. This led to the examination of optional enrichments separately from the essential meaning. A useful structure for analysing enrichments has four headings:

- inclusion,
- dependencies,
- hiding,
- protection.

In systems engineering discourse, all four need to be addressed explicitly in order to provide a full and clear notion of what encapsulation choices are possible. These four provide the cornerstones for a system's encapsulation strategy. They are discussed in turn in sections 6.2.1-6.2.4. Section 6.2.5 warns that not all encapsulations can be judged by criteria applied to module design.

6.2.1 Inclusion principles

In the context of encapsulation, inclusion principles are those principles which determine which items are inside the encapsulation and which are not. These principles should be determinate and unambiguous with respect to four aspects of what is enclosed — an encapsulation's boundary, scope, content (both in terms of its nature and its internal structure) and permeability.

6.2.1.1   Boundary

At the very least, an encapsulation must have a boundary that is well enough defined to make clear what it encloses and what it does not. From the boundary definition, it should be possible to determine for every item which could be enclosed whether it is in the encapsulation or is not. This means three situations should not be allowed to persist:

- not knowing whether particular items are in or out of the encapsulation — e.g. the conditions for being in (or out) are not clear enough for us to make the decision or there is too little relevant information about the item to decide;

- having some items which are both in and out of the encapsulation — e.g they may satisfy the conditions for being in the enclosure and also satisfy the conditions for not being in the enclosure;

- having some items which are neither in nor out — e.g. they fail to satisfy the conditions for being in and fail to satisfy the conditions for being out.

When the inclusion principles allow relatively complicated encapsulations it may be difficult to treat the encapsulation as a whole or even to know what has been encapsulated.

Heterogeneous encapsulations enclose different kinds of items. Objects which encapsulate data and processes as attributes and methods, respectively, are an example. The inclusion principles for the different kinds of items are likely to be different and all need to be clearly expressed. It may be that for some systems (e.g. those based on objects but implemented in non-OO languages) all kinds of items can only be encapsulated at some levels (e.g. at the conceptual and logical levels) but not at others (e.g. implementation or physical levels). This needs to be recognized and its implication, that some object technology benefits can not be realized, must be understood.

Homogeneous encapsulations, i.e. ones which contain only one kind of item (e.g. only data items or processes), also can pose problems.

Difficulties arise when either heterogeneous or homogeneous items are not atomic but are made up of parts. Kim [7], for example, distinguishes between simple and composite objects — a simple object being one whose attribute values are not references to other objects, and a composite object being one whose attributes may have values which are references to other objects. Figure 6.1 shows both a simple and a composite object. The simple object has the name of the manufacturer as a value in the object itself. The composite object has as its value for manufacturer a reference to another object — a company.

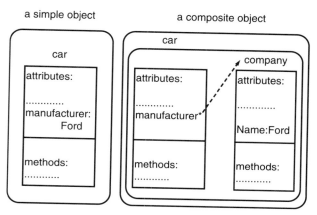

Fig. 6.1   Simple versus composite objects.

Encapsulations which enclose simple objects are relatively straightforward once the issue of encapsulating data and process has been addressed for objects in general. Encapsulations involving composite objects are more complicated, however, and a policy for their encapsulation needs careful consideration.

Suppose an encapsulation — such as a car fleet — is defined to allow a collection of cars to be regarded as a single object. The fleet can have attributes such as the company which owns it, its monetary value, its manager, and its size. Its behaviour might include conforming to company policies, generating insurance costs, and decreasing in value each year.

There is no problem in knowing what is encapsulated in the fleet if cars are simple objects such as the one in Fig. 6.1 — at least conceptually the entire car object is in the fleet. There may be a problem, however, if cars are represented as the composite objects in Fig. 6.1. Then the object representing the manufacturer is part of the car object. Decisions need to be taken about the company objects, the car manufacturers. Are the car manufacturers to be regarded as encapsulated in the fleet as part of the car objects or not?

There are software design implications with either choice. If they are included, their life cycles may be tied to that of the fleet — an issue discussed under the heading of existential dependency in section 6.2.2 below — and it may be decided to keep the code relating to them with that of the fleet and to physically store them with the fleet. If they are not encapsulated, it might make it difficult to treat the fleet as a single whole thing. There may be stronger arguments for keeping their code and physical storage apart from that of the fleet with all the design ramifications of those decisions. The details of this kind of encapsulation need clear definition and careful thought.

6.2.1.2   Scope

The scope of encapsulation may be based on any one of a number of determinants — content (e.g. a single function, transaction or real world thing), an architectural feature of the program (e.g. module, division or section of code), the physical structure of the system (e.g. a file or a DB2 table), or be related to some aspect of processing (e.g. the platform on which it occurs or the site at which it takes place). Whichever is chosen the determinant should be made explicit.

Encapsulations can also differ widely in the extent of what they encompass. Systems themselves are a kind of encapsulation as are the sub-systems, modules, data units and functions of which systems are made. Systems based on object decomposition and the notion of classes will have objects, classes, attributes and methods as kinds of encapsulations. They can all provide well-defined boundaries, scoping and content.

To get the maximum benefits of object technology, however, the encapsulations must have object-like qualities, clean interfaces and the encapsulation of data and operations, which not all of these kinds of encapsulation possess. This is discussed further in section 6.2.5 below.

6.2.1.3   Content

In systems based on objects, data belonging to an object and the operations on that data (discussed above) should be in the same encapsulation. This may not be possible in the code because of language constraints, or in the physical system because of platform or other hardware constraints. Nonetheless, the logical encapsulation of these parts into objects should be expressed explicitly in analysis and design documents.

The content of one encapsulation may also be other encapsulations (e.g. an object may encapsulate other objects; a module may encapsulate other modules and/or objects). Where this happens a number of points must be made clear:

- the kind of encapsulations that are able to become enclosed items (there may be no restrictions or severe constraints);

- the relationships that exist among the enclosed items;

- whether items can be more than one encapsulation;

- the relationship between the internal items and the encapsulation which encloses them.

The last two points require further elaboration. Point three is discussed here. Point four is dealt with in section 6.2.2.

There are at least two ways for items in one encapsulation to also be in another encapsulation — when the encapsulations are nested and when they overlap (see Fig. 6.2).

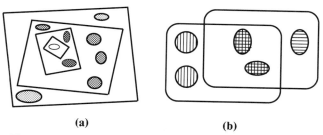

(a)   **(b)**

Fig. 6.2   Encapsulations — (a) nested and (b) overlapping.

For an encapsulation to be called nested all the items in an internal encapsulation must be enclosed by a 'larger' encapsulation, as in Fig. 6.2(a). Overlapping encapsulations, such as in Fig. 6.2(b), each have items which are not enclosed by the other encapsulation and so are not nested. Where overlapping partial encapsulation of an item is undefined, items of the kind shown by the double cross hatched items in Fig. 6.2(b) are not allowed. These notions of overlapping and nested encapsulations may be useful in reasoning about containment and sharing discussed below.

6.2.1.4   Permeability

Another issue related to inclusion is whether the items can move between encapsulations during their lifetime in the system, i.e. whether once included always included, or not. This is related to the permeability of the 'walls' of the enclosure. If the walls are impermeable, no item that was inside the encapsulation can later be outside it and no item that is outside can 'move' in. If the 'walls' are permeable, then there can be a 'flow' of items between inside and outside. It may be useful to define enclosures whose 'walls' are semi-permeable, permitting perhaps only one-way flow or flow of only certain kinds of items across the boundary.

Illustrations can be found for all these options. For example (discounting destructive activities like tearing and overpainting), a photograph impermeably encapsulates the people in it — no person can be added or removed from it. Most parties are a permeable encapsulation of guests — they can enter and leave the encapsulation. A roster of former mayors is a semi-permeable encapsulation of people — no person once included can later be excluded, but each new person serving in that role is added to the encapsulation.

6.2.2   Dependencies

Relationships between encapsulations can be defined by the structures in which they are found, how they depend on other objects and what can be shared between them.

6.2.2.1   External structures

Objects and the classes of which they are instances are usually in one or more hierarchies such as generalization-specification hierarchies and part-whole hierarchies. In terms of encapsulation, the implication of being part of any such structure depends on how the structure is defined. How much an encapsulation that is low in the structure can 'see' of encapsulations above it must be defined (see section 6.2.5 below). What the structure implies in terms of dependencies must also be defined. Two kinds of dependencies have been identified in this study as particularly important — existential dependency and state dependency.

6.2.2.2   Existential dependency

Existential dependency is concerned with how the existence of one item depends on the existence of other items. It is important to distinguish between dependent and non-dependent items, to consider sharing of items, and to think about conditions which affect the existence of encapsulations and the items they encapsulate.

Dependent versus non-dependent

Returning to the example shown in Fig. 6.1, the system may have been designed so that the car manufacturer 'company' is only used as part of the composite item 'car' and is meaningless to the system unless it is associated with 'car'. In this case, the system is also likely to have been designed so that 'company' ceases to exist in the system when 'car' ceases to exist — 'company' is existentially dependent on 'car' in such a design.

An item and all its dependents should be treated as a whole, both within the system for which they were designed and when reused, i.e. there should be some form of encapsulation whose scope includes them all.

The system may, however, be designed so the car manufacturer 'company' continues to exist even when the 'car' for which it is the manufacturer ceases to exist in the system. In this case the one, 'company', is associated with but is not a part of the other, 'car'. This might happen, say, because 'company' is the manufacturer for several 'cars' (i.e. they all reference it — it can be regarded as shared by them), or because it is important for other purposes, e.g. it is also a client or the system needs to keep it for historical purposes. In these cases 'company' is not likely to be part of 'car' but a separate item with an independent existence. It is not existentially dependent on 'car'. This can be represented as in Fig. 6.3, where the absence should be noted of the rounded box which in Fig. 6.1 enclosed the 'company' object with the rest of the details of 'car'.

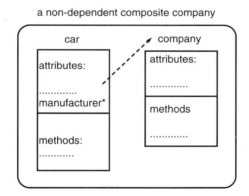

Fig. 6.3   An example of an association between objects.

In the case of non-dependent items, all individual items may be regarded as separate and may be reused individually where appropriate.

Sharing policies

An item which is dependent on another item for its continued existence should not be shared (i.e. also referenced by other items). Otherwise the shared item could cease to exist for reasons which have nothing to do with some of the independent items referencing it. Its demise would leave those items with nothing to reference and probably no way of 'knowing' this was the situation. Other sharing policies arise from state dependencies discussed below.

Conditions for continued existence

The issue of how long items remain in encapsulations was touched upon earlier, in the context of the permeability of an encapsulation. It is also important to consider whether items can survive the demise of encapsulations in which they are enclosed and whether an encapsulation can survive the demise of one or more of the items it encapsulates.

One design possibility is existential dependence between encapsulations and items it enclosed. A strict version of this would mean that the encapsulation is impermeable and lasts only as long as all the items it encapsulates exist. This is what usually pertains for simple objects [8]. Each object, including all its attributes and methods, is created and deleted as a whole. The values in its attributes may change while the system is running. It is not usually the case, though it is not inconceivable, that the internal structure of an object (i.e. the number and specification of its attributes, the number and implementation of its methods) changes as a result of running the system.

It is worth considering whether this 'all or nothing' strategy should be adopted for other encapsulations such as islands (see section 6.2.4 below) and containers (such as collection in Smalltalk). The answer may well be no. The flexibility of being able to add or delete an item from the system and/or an encapsulation independent of the continued existence of the encapsulation may be vital to the design. For example, cars could come into the system and not join the car fleet, e.g. cars hired for special occasions. Also a car may be regarded as being out of the fleet once it has been replaced, but remain in the system until it is sold. Also the fleet as such may cease to exist, but the cars need to remain in the system associated with individual employees who might require parking permits or claim expenses for using them for business. These situations require existential independence between items and encapsulations in which they may be enclosed.

6.2.2.3   State dependency

State dependency is about how the state of an item (the values of its attributes) affects its interactions and relationships with other items in the system, including encapsulations.

The danger that an encapsulation may cease to exist without other items, which reference that encapsulation, 'knowing' of the change was mentioned in the discussion of existential dependency above. A change in the state of an item can also have unwelcome side effects. It is a problem which is highlighted by both Yelland [9] — in the context of experiments to test his hypothesis that encapsulation (he uses an enriched version of the notion which includes notions of hiding and protection) is frequently and unwittingly breached even in as pure an OO language as Smalltalk — and Hogg [10], in the context of the general problem of aliasing (i.e. having more than one 'path' to an object). It is a danger which exists but is one which is less pervasive than one might imagine from reading Yelland's and Hogg's work.

Safe changes of state

Items with only a single reference to them, unaliased items, are not a source of danger, because by definition there is no sharing. Immutable items, ones whose state cannot change, are not a source of danger even if there are several items referencing them because by definition there is no change of state — Yelland [9] excludes immutable objects such as integers, from his experiments for just this reason.

That leaves multiple references to mutable objects, ones whose state can change, as the potential source of state dependency problems, and not all of these cause difficulties. For some, changes in the state of objects they reference pose no threat, so long as the referenced object remains 'correct'. Here 'correct' means that all of the attributes of the referenced object have values which are legitimate within the definition/specification of the object, that it, as a whole, satisfies any constraints or conditions which are part of its specification, and they accurately reflect the aspect of the real world or system they are intended to represent.

It is all right, for example, for a 'car' item to be referenced by an item which is its owner and another which is its driver and another which is the company which underwrites its insurance. The fact that some aspect of its state, e.g. its location, changes does not in general threaten the objects which reference it. In fact it may be important that the value of its location attribute is allowed to change so that the system in which it is an item continues to be an accurate reflection of the real world situation.

In general, suppose an item — say D — is defined in terms of a relationship between items A, B and C (the number of items is not significant). If A, B and C are mutable there may be problems. If A, B and C can change in ways which do not affect the relationship which was used in defining D, then there is no problem. If they can change in ways which do affect that relationship then there may be problems.

For example, suppose D in the example above was a cumulative ordered list of names of employees of a firm and that each employee was represented in the system as an item in its own right. A, B and C from the example above would each be an employee.

If the list of employees were put in a sequence according to the date the person first joined the firm, an immutable attribute, there might be problems in ordering the list because of ties — two people joining on the same date

— but once ordered the list would remain correct even though the state of employees changed in other ways. The list would only become 'incorrect' if a new person joined the firm and the list were not altered to reflect this, i.e. it would be incomplete. This is not an incorrectness caused by a change of state of an employee on the list.

Dangerous changes of state

If, instead, the list were ordered by surname of the employee, a mutable attribute, and one or more of the employees changed their surnames (e.g. through marriage or deed poll), the list would become incorrect. It would be incorrect because the ordering relationship among the employees was no longer correct. The employee items are not aware of the sequence the ordered list expects because that relationship 'belongs' to the ordered list item. The correctness of the ordered list does not depend on its relationships with individual employees, ones over which it might have some control, but their relationship to one another under the relevant sequencing scheme. Thus changes in items it references, i.e. the employees, can make the list incorrect. It is a change of state on an attribute used to define a relationship between items with that attribute which is dangerous.

Unless software engineers identify the dependency of one item on a relationship between other items, adverse side effects arising from them are difficult to avoid or detect. For complete protection, an item such as the ordered list would need to know when the value in any of the items involved in that relationship changed on the attribute (e.g. the surname of the employee) which was crucial to the relationship upon which it depended. One mechanism would be to poll those items constantly, checking for change of state on the relevant attribute. Another mechanism would be to have the ability to instruct those items to notify the ordered list when the relevant attribute changed. Either mechanism complicates the system, with all that means for additional design, implementation, testing, and maintenance costs — both are also likely to diminish performance.

There is a strong case for identifying the kinds of encapsulations for which sharing or other dependencies can be dangerous and for preparing guidelines for developers about using them. This, and the development of associated metrics, is an avenue recommended for future work.

6.2.2.4 Communication dependencies

Communication dependencies are about the need for one item to interact with another item, say, through sending messages. A typical requirement in an object technology system is that the message sender does not need to know internal details of the target of its message for the message to work effectively, but it must be aware that the target exists and know enough about its public interface to be able to send an appropriate message. This is similar to module design. The design issues relate to what details of the target item the message sender needs to know and whether it accesses the internals of the target item directly. These are covered in sections 6.2.3 and 6.2.4 below.

6.2.3 Hiding — opacity of the walls

Opacity is about what an 'observer' on one side of the enclosure wall can 'see' of what is on the other side. It is about visibility and information hiding as distinct from accessibility which is discussed in the next section. It raises the question of what it means to see parts of a system. Here 'seeing' an encapsulation is distinguished from 'seeing inside' the encapsulation.

'Seeing' the encapsulation will be taken to mean:

- 'being aware' that the encapsulation exists;
- 'knowing' what is in at least some part of its interface.

'Seeing inside' the encapsulation means being aware of the existence and the nature of the items in the encapsulation. To be of use, everything in the system must 'see' or 'be seen' by some other item in the system, but they do not need to allow other items to 'see inside' them.

Objects can be, and often are, developed so that the only way to use the services they provide or to access the information associated with them (their state) is via their interface, usually by sending them a message. Under such a regime information is hidden in objects. They can 'be seen' but not 'seen into', making it possible for their internal data structures to be redesigned and their methods reimplemented without affecting the rest of the system.

6.2.3.1 'Seeing into' an encapsulation — information hiding

Instances of subclasses or derived classes usually 'see into' their parent, but not always. For example, it is possible to program in Smalltalk [11] so that an instance of a subclass interacts with its parent by passing messages to its interface rather than by directly accessing its internals, the more usual programming style.

C++ allows internal details of a class to be public, protected or private. Instances of the class itself can 'see' and (to anticipate the next section) manipulate all of its own parts. Instances of its subclasses can 'see' its protected and public parts, but not the private ones. Instances of classes which are its clients can 'see' its public parts only. However it also has the notion of friends. Instances of classes designated as its friends can also 'see' and manipulate all of its internal parts. The language allows but does not enforce the notion of an interface and message passing, so the distinction between 'seeing' and 'seeing into' is always useful. Confusion about the distinction between 'seeing' and 'seeing into' arises wherever the implementation language does not support the notion of interface.

The issue also arises because many designers specify interfaces which closely mirror, and include details about, the internals of the object. This makes it look as though external objects are 'seeing into' an object (although this may not be the case). It also makes it apparently more difficult for future maintainers and redevelopers of the system to make changes to the object's internals.

Not all kinds of encapsulations, not even all objects, are designed within information-hiding interfaces. There are development choices to be made about both 'seeing in' — the perspective of a part of the system external to the encapsulation — and 'seeing out' — the perspective of an item inside the encapsulation.

Related to these but not strictly a matter of the opacity of the wall of the encapsulation is its own perspective, i.e. the perspective of the encapsulation itself — what it can 'see' of its internals and what it can 'see' of its environment. In so far as the encapsulation can, as a whole, be thought of as seeing anything, it often has sight of all its internals, at least of their existence as its internals. Whether it has sight of their nature is another question, especially if its internals are themselves encapsulations. These issues are not explored further here, but remain on the encapsulation research agenda.

6.2.3.2   Useful range of opacity

The full range of opacity, from opaque through to transparent, can be useful in system development. The more opaque the walls, the more freedom there is to alter the internals of an encapsulation without affecting the rest of the system — the more transparent, the less freedom to change.

At one extreme, it is possible to have an interface to an encapsulation which bears so little relation to what is actually enclosed as to be completely opaque. The telephone system is an encapsulation which comes close to being opaque to its users. The handset, with its earpiece, number-pad and ringing device, provides functionality but virtually no visibility into the system to which it is an interface. A C++ class with all its parts declared as public would be transparent.

6.2.4   Protection from direct alteration — islands

The accessibility of an encapsulation's internals to manipulation by parts of the system outside the encapsulation is related to, but is not the same as, opacity. 'Seeing into' an item does not necessarily allow what is seen to be altered. It may involve read-only access. In many object technology systems other parts of the system have no direct access to the internals of an object. Instead they must send messages to the object requesting whatever they want — information, to update the object, or use of one of its services. The object responds to each message by executing one of its methods. Its own methods access and manipulate its data.

Hogg's notion of islands [10] is an attempt to control access to collections of objects. One object serves as the bridge to the island and is the only point of access to those objects which make up the island. No object which is outside the island can hold a reference to any object which is part of the island other than the bridge object. Objects which are part of the island can hold references to one another and to the bridge. The bridge is 'aware' of all changes to the state of the island, so, presumably, it would be able to prevent adverse side effects arising from changes in state, but only if its design permitted it to do so. Relationships, discussed in the context of state dependencies above, would still need to be identified and the design address their management. Hogg suggests that container structures, such as the ordered list, are natural candidates to be treated as islands.

6.2.5   Non-interchangeability between kinds of encapsulation

The term object is increasingly being applied to any encapsulation, but not all encapsulations provide the full benefits of object technology. It is important to keep the distinction between those which encapsulate data and processes and can, therefore, be regarded as objects, and those that do not and should not be regarded as objects. What constitutes good design may differ between different kinds of encapsulation as well, e.g. differences between modules and objects.

- Modules — 'packages' or 'tasks' in Ada, 'files' in C++, 'units' in Object Pascal — provide well-defined boundaries between parts of a program. They usually reflect the functional decomposition of the system, define the physical structure of the program and may define the physical architecture of the system. They are usually designed to have low coupling (indicated by little dependence and little communication between modules) and high internal cohesion (indicated by a high level of interdependence and intercommunication between elements inside the module) (see Buxton and McDermid [12] for further discussion of modular cohesion and coupling). Objects have similarities but are also somewhat different.

- Objects, and the classes of which they are instances, reflect the object decomposition of the system, not necessarily its functional decomposition. They define the logical structure of a system, but may not map on to the physical divisions of the program or the physical architecture of the system. For example, in C++ the header file (e.g. with .h extensions) and code file (e.g. with .c extensions) for the same object may be physically separated in the program; in OO database systems the data part of the object may be stored separately from the methods of the object (the operations on its data).

Two principles of good object technology design — clear object-based division of responsibility and 'crisp' objects (see below) which together make for good object definition — are similar to the two principles of good modular design, but are not exactly the same and have different, and in some respects contrary, indicators.

A crisp object is one whose class definition ensures that all the data relevant to the object and all the operations on that data are encapsulated in the object, and that all extraneous data and operations on that extraneous data are excluded. In some respects this promotes both high coupling and low cohesion, the opposite of what is sought in the case of modules.

Objects are designed to carry out their responsibilities by accessing and operating on their own data as necessary and interacting with other objects which provide services through accessing and operating on their data. Extensive communication between objects, e.g. high coupling, is usually tolerated.

Objects are also designed so that the items (attribute values and methods) they enclose relate to a single real-world thing, not because they relate to one another. A date, for example, in one object will be significant to the object in which it is enclosed, but could also be intimately related to other date objects. In some important ways, e.g. in how it is stored and displayed, it is more closely related to them than to other items, such as names, in its encapsulating object. This could be interpreted as meaning that objects have low cohesion.

The point is that the indicators for good module design must be examined carefully before being applied to objects. Indiscriminant labelling of encapsulations as objects is inappropriate and may be misleading. For encapsulations to provide the benefits of object technology they must have those characteristics of objects, for it is they that deliver those benefits.

6.3   EXAMPLE FROM A BT SYSTEM

This section considers encapsulations used in COSMOSS (computer orientated system for the management of special services), a BT system currently under development which is discussed in some detail in Chapter 5. It is a particularly interesting example because, as it is not implemented in an OO language, there was no default object technology strategy that could be adopted. The development team has had to work within strict constraints but have been left with some interesting decisions.

Unable to use many of the features of the object technology approach — e.g. inheritance, and polymorphism — because of having to implement the system in COBOL, the development team hoped to get some object technology benefits by making use of encapsulation. Their encapsulation strategy is discussed in this section in terms of features highlighted by unpacking the four cornerstones of encapsulation introduced above:

• inclusion principles — such as the boundary of encapsulations, their nature, content and scope;

• dependency issue — structures, sharing, duration and permeability;

• hiding mechanisms — opacity, visibility and information hiding;

• protection choices — accessibility of encapsulation to external manipulation.

For COSMOSS, the development team has opted for a flat architecture with active transaction-controlling 'objects', Transactions, which drive passive 'objects' called Capsules or Super Capsules (see Fig. 5.1 in Chapter 5). Capsules and Super Capsules access and operate upon data in tables and super-tables respectively, all of which are discussed below. Thus COSMOSS has five kinds of encapsulations — Transactions, Capsules, Super Capsules, DB2 tables, and super-tables[1].

The first three may be either COBOL modules or programs and are made up of COBOL code. Capsules / Super Capsules are made up of code which operates on one/several (respectively) DB2 tables. A Capsule encapsulates the operations associated with a data entity (a single DB2 table). A Super Capsule encapsulates the operations associated with a super-table. A Transaction calls operations in one or more Capsules / Super Capsules using

---

[1] A DB2 table is a third-normal-form relational database table. A super-table is a table which results from the merging of several DB2 tables, a relational database 'join' command.

a predefined set of parameters for each operation and expecting a predefined set of outputs from that operation. The discussion below focuses on the data aspects of the system, the Capsules and Super Capsules.

6.3.1   Capsules and Super Capsules

For the COSMOSS design a data entity is taken to be one DB2 table. A Capsule is the code associated with one data entity and as such is a COBOL module or program rather than an object in the full sense of the term. There is one Capsule for each DB2 table. The Capsule contains all the code necessary to retrieve the current state of the data entity and for the support of its full life history. A Super Capsule contains the code for creating joins of related tables and retrieving the state of related tables. However, they cannot modify tables [13].

6.3.1.1   Enclosure

Physically, data and operations on that data are separate. The data is in the DB2 database or temporarily held as super-tables. Code is in Capsules and Super Capsules.

Conceptually, for analysis and perhaps even for high level design, it is appealing to regard Capsules and Super Capsules as objects, enclosing both data and operations. There is, after all, a one-to-one relationship (discussed below in section 6.3.1.2) between Capsules and tables, and between Super Capsules and super-tables. However, as COSMOSS is presently designed, Capsules and Super Capsules should not be regarded as encapsulating data with operations. This is not because there are different physical locations for data and operations as there might well be, but because the Capsule is not the sole interface to a table — Super Capsules can also read DB2 tables directly when they do their joins — and the data on which Super Capsules operate can only be updated through Capsules.

If Super Capsules had been designed to call Capsules to retrieve the required data, then the Capsule would have been the sole interface to the table and it would have been more appropriate to regard it as an object. But designing it that way would have introduced a level of referencing which COSMOSS designers wanted to avoid. In fact, the major decision, to adopt a flat architecture, was based on their policy to minimize the use of referencing in the system.

Another reason for regarding data as separate from operations in COSMOSS is that a change in the structure or design of a DB2 table (e.g. the number or content of the columns it contains) potentially affects the code not only in its Capsule but also in all the Super Capsules which use the table in creating joins. Thus a table could be 'in' a number of encapsulations, but no syntax has been defined for such 'overlapping' encapsulations.

6.3.1.2 Content and scope

At the Capsule level, the system has been designed so that all the code in the Capsule relates to the one DB2 table only, and all the code needed to support and retrieve that table on its own is in the one Capsule, i.e. there is a one-to-one mapping between Capsules and DB2 tables. There will be as many Capsules in COSMOSS as there are DB2 tables in the database. The scope of each Capsule, then, is all the operations affecting a single DB2 table, or 'the table' for short. There is a one-to-one mapping between the 'super-table' created by a join and the Super Capsule that creates that join, so the scope of the Super Capsule would be 'the join'.

The on-line data model used in the project was not the final COSMOSS model. The scope of DB2 tables was still being defined, but they are likely to be 'less than' the data part of an object. They are likely to represent some aspect of the object rather than all aspects. The scope of a super-table is likely to be closer to the data part of a whole object, but that also remains to be seen.

On the scope dimension, the success of this strategy in providing the full benefits of an object approach depends on the DB2 tables making sense as 'objects'. If they do not, then the strategy can provide only some advantages of object technology, for example:

- design and development benefits because the code can be unambiguously partitioned and Capsules and Super Capsules all developed independently of one another;
- easier maintenance and testing because of the one-to-one mapping;
- possible reuse of some parts of each Capsule / Super Capsule by other applications using the same database.

The benefits of conceptual clarity and coherence that come from having the system reflect the real world, stability that comes from object decomposition, and performance gains that come from obviating the need for joins, would all be lost.

The fact that the design includes the notion of 'Super Capsules' is an indication that the DB2 tables do not make sense as objects, that joins will be required, and that some benefits from object technology have been lost, hopefully traded-off to gain other kinds of benefits.

6.3.1.3   Structures

As far as one could tell, neither Capsule nor Super Capsule design depended on rich internal structures. Their code will presumably reflect the usual structures of COBOL such as 'section' and 'paragraph' and 'CASE' structures mentioned in Chapter 5 and may be written following the code template recommended there (see Table 5.1 in Chapter 5), but that is not part of their design.

Neither Capsules nor Super Capsules form part of any external structure in the sense discussed in section 6.2.2 above. They are simply elements of the flat architecture adopted in COSMOSS to minimize reference chains and simplify the design. In particular, Capsules are at the same level as Super Capsules, not parts of Super Capsules nor 'beneath' them in an inheritance structure.

This, plus the fact that Super Capsules have direct access to tables to retrieve their state, means that code for reading tables involved in joins will appear in at least two and possibly more places — in the Capsule associated with the table and in the Super Capsule which creates joins in which it is involved. The fact that a Capsule / Super Capsule can be called by different Transactions (see below) means that there can be extensive reuse of code in those encapsulations. Taking these two together means that the COSMOSS design reduces repetition of code but does not eliminate it.

Tables have internal structure of rows, columns and cells which hold values. They are part of flat external structures where tables are connected by links represented (e.g. by pointers and keys) in the usual DB2 way.

6.3.1.4   Dependencies and sharing

Transactions can share Capsules and Super Capsules and the operations in them — this means that a Transaction can call some or all (or none) of the Capsules / Super Capsules called by other Transactions. They may call the same or different operations from a shared Capsule / Super Capsule. This means that changes in the operations in Capsules / Super Capsules potentially affect several Transactions. This is discussed in more detail below in sections 6.3.1.6 and 6.3.1.7.

Unless there is indirection in the tables (e.g. the cell of a table containing a reference to another table or cell(s) of another table) there are no dependencies in COSMOSS of the kind discussed in section 6.2.2.

Tables, however, are shared between Capsules and Super Capsules in that both kinds of modules may retrieve the state of the table (i.e. read all or part of it and pass what is read on to the Transaction).

6.3.1.5   Duration and permeability

Both Capsules and Super Capsules in COSMOSS have a duration which might be termed 'permanent' in that they are unchanged by the operation of the system. They are only altered by rewriting them during maintenance, redevelopment or enhancement of the system. It also follows that they are impermeable, i.e. no operations move into or out of them.

If, as seems the case, DB2 tables are neither created nor deleted as a result of operating the system (but only during maintenance, redevelopment or redesign), then they are also permanent. They are impermeable at what might be called the class level, in that neither the nature nor number of columns (hence the attributes they represent) is changed while the system is operational. They are permeable, though, at what might be termed the instance level, as they can have rows added, altered or deleted (increasing, changing or decreasing the instances they represent).

Super-tables created by Super Capsules are not saved in the system, and so may be termed 'transient'. They use snapshots of the tables from which they are made and once created do not change and so are impermeable.

6.3.1.6   Opacity — visibility — information hiding

A Transaction calls a Capsule / Super Capsule using a parameter set which identifies which opeation(s) it wants that Capsule / Super Capsule to carry out. In this sense the operations relevant to the Transaction are 'visible' to it. It 'knows' what parameter set is appropriate to address which Capsule / Super Capsule. As long as the parameter set and set of outputs it is expecting are not changed, a developer/maintainer can alter the code in the Capsule / Super Capsule code without affecting the code of the Transaction.

Thus Capsules / Super Capsules hide information[1] while providing full visibility to Transactions. Transactions have 'what' visibility not 'how' visibility. Capsules / Super Capsules have no visibility of any other Capsule / Super Capsule.

When appropriate, a Capsule / Super Capsule can see into its DB2 table(s) and can pass values it reads to the Transaction. Everything in a table is visible. No changes can be made to the structure or content of tables without, potentially, affecting subsequent redevelopment/maintenance. Unless there are references between tables, they cannot really be said to have visibility of each other. Nor do they 'see' any other encapsulation and so may be termed 'blind'.

---

[1] The information is not data but the code which implements the operations provided by the Capsule / Super Capsule.

6.3.1.7   Accessibility to alteration

Nothing in the COSMOSS system can alter the code in Capsules and Super Capsules, so they are regarded as inaccessible, and therefore, in this sense, fully protected.

Only the table's associated Capsule — no Super Capsule, no Transaction, and no other table — can alter (i.e. update by adding rows, deleting rows, or altering the contents of cells) a DB2 table. So, although a DB2 table may be shared by different 'objects', endangering the system is avoided. A danger which is not avoided by this mechanism alone, but perhaps is avoided in some other way in the design (e.g. by concurrency and commit strategies), is that a Super Capsule is not 'aware' (it is not notified) of crucial changes in the table. Presumably the code of a Super Capsule includes a check on any aspect of the state of the table which influences the way they create their joins. This has been catered for by allowing Capsules and Super Capsules to read tables and use the result in their operations as well as or instead of passing that data to the Transaction.

6.3.2   Summary of COSMOSS encapsulation analysis

The COSMOSS system has five kinds of encapsulations. Three of them — Transactions, Capsules and Super Capsules — are COBOL programs or modules encapsulating operations. Their scopes are a transaction, a table and a join respectively. None are involved in external structures. Their only internal structure is that dictated by COBOL. Several Transactions can share operations of Capsules / Super Capsules. All of these code encapsulations are permanent and impermeable. Transactions 'have knowledge' of operations provided by Capsules / Super Capsules, but cannot 'see' how those operations will be performed. Capsules / Super Capsules have full visibility of tables. Only the associated Capsule/ Super Capsule can access a DB2 table/super table, respectively.

The other two of them — DB2 tables and super tables — encapsulate data. The scope of DB2 tables is likely to be an aspect of an object. The scope of a super table is the join which is likely to correspond more closely to the data part of all of an object. Tables have an internal structure of rows, columns and cells which hold values. They are part of flat external structures in the usual DB2 way. A table may be shared between its Capsule and several Super Capsules, but only the Capsule can update. There is no danger to the Capsule from unexpected changes of state of its table. Super Capsules may be in danger from unexpected changes of state of one or more of the tables it uses to make its join. For its creation, a super table is dependent on the

DB2 tables it requires for its join. DB2 tables are impermeable at the level of their internal structure (as class), but permeable in terms of the number of entities they represent (as collections of instances). Super tables are transient and impermeable. Tables are fully visible but blind.

6.4   CONCLUSIONS

A system can involve numerous kinds of encapsulations including modules, classes/objects and islands. These can be arranged in a flat structure or be in several layers, and be nested. No syntax for overlapping encapsulations has been defined here, but these are also conceptually possible.

Developers should make their encapsulation strategy explicit (in as many as possible of the framework areas discussed in this chapter):

- inclusion principles — unambiguous definition of what is inside and what is outside an encapsulation, its content, scope, kinds of encapsulations used (e.g. modules, classes/objects, islands, containers), what is the nature of what they enclose, how much they include;

- dependencies — including details of their internal and external structures, existential, state and communication dependencies, sharing policies, their duration and permeability including whether, under what circumstances and with what effects items may move between encapsulations;

- hiding — what can be 'seen' and changed without affecting other parts of the system, what can be 'seen into' and therefore cannot be changed without affecting other parts of the system;

- protection from direct external alteration — which, if any, parts of the system outside the encapsulation can directly alter some (if so, which) or all of what is enclosed.

Further work identifying the kinds of encapsulations for which sharing is dangerous, under what circumstances it is dangerous and guide-lines for eliminating or minimizing the danger are still needed.

Visibility is crucial to systems. Encapsulations must see or be seen by something if they are to be effective parts of the systems. But the more an encapsulation can be 'seen into', the less freedom there is for the internals of the encapsulation to be altered without affecting other parts of the system.

Metrics are needed for the effects of referencing and for design, implementation, testing and the maintenance effort required for different encapsulation strategies.

The BT COSMOSS project is an interesting development to analyse in the context of encapsulation. The system is constrained to being implemented in COBOL which limits the object technology features it can employ and makes any object technology design choices those of the developers not of an OO programming language. The simplicity of its flat architecture has many advantages, especially the reduction of referencing. Capsules and Super Capsules, two of the five kinds of encapsulation used in the system, allow extensive reuse of code. The fact that Super Capsules can access DB2 tables directly means that there is some repetition of code. It also means that any restructuring or redesign of the tables has more impact than would have been the case if Super Capsules were constrained to retrieve the state of tables from the appropriate Capsules.

Encapsulation is an important and powerful feature of the object technology. It can be harnessed to make large legacy systems more flexible and more future-proof. This chapter has provided a four-area framework for analysing encapsulation choices. It is an aid to making the strategy adopted more explicit and less ambiguous. Both are required if much needed metrics relating to encapsulation are to be developed.

REFERENCES

1.   Bloor R: 'The object management guide', Butler Bloor Ltd, Milton Keynes (1992).

2.   Booch G: 'Object oriented design with applications', The Benjamin Cummings Publication Co Inc, Redwood City, CA (1991).

3.   Coad P and Yourdon E: 'Object oriented design', (Yourdon Press) Prentice-Hall International, London (1991).

4.   Rumbaugh J, Blaha M, Premerlani W, Eddy F and Lorensen W: 'Object oriented modelling and design', Prentice-Hall International, London (1991).

5.   Jeffocate J, Hales K and Downes V: 'Object oriented systems: the commercial benefits', Ovum Ltd, London (1989).

6.   Berard E V: 'Abstraction, encapsulation, and information hiding', Berard Software Engineering, Inc., 101 Lake Forest Blvd., Suite 360, Baithersburg, Maryland 20877, USA (1991).

7.   Kim W: 'Object oriented databases: definition and research direction', IEEE Transactions on Knowledge and Data Engineering, $\underline{2}$, No 3, pp 327—341 (September 1990).

8.   Cordingley E: 'Simple gardening with objects', Internal BT report (1992).

9.   Yelland P: 'Models of modularity: a study of object oriented programming', PhD Dissertation, University of Cambridge (1991).

10. Hogg J: 'Islands: aliasing protection in object oriented languages', OOPSLA '91, pp 271—285 (1991).

11. Goldberg A and Robson D: 'Smalltalk-80: the language', Addison-Wesley (1989).

12. Buxton J and McDermid J: 'Architectural design', Chapter 17 in McDermid J (Ed): 'Software Engineer's Reference Book', Butterworth and Heinemann, Oxford (1991).

13. Shearer D: 'Object orientation in COSMOSS development', Internal BT report (1992).

7

FORMAL METHODS IN OBJECT ORIENTED ANALYSIS

J C R Wilson

7.1   INTRODUCTION

Within BT there is a need to be able to carry out requirements analyses of systems which are large yet which are subject to change during their lifetime, for example the network routeing management system described in Chapter 13.

The requirements specification of such a system needs to be unambiguous yet at the same time simple in structure and easily understandable by people from a wide range of specialisms. Because the system will be subject to change, the specification also needs to be easily modifiable and extendible.

In this chapter an approach is suggested which combines the precision of mathematics, the simplicity and clarity of diagrams similar to those found in traditional structured analysis and the flexibility and reusability offered by object orientation.

Traditional engineering mathematics was based on calculus but the mathematics used here is based on set-theory and is in many ways far simpler[1]. It will be used to specify with great precision how we want the basic building blocks of our system to behave. In fact we will use a notation called Z[1], which was developed for applying set-theory to the specification

---

[1] The basic concepts of set-theory have formed a standard part of many secondary school mathematics syllabuses since the 1960s.

of software systems, together with a few minor additions so that it can be used in an object oriented context.

There are already a number of techniques for object oriented analysis available [2-8]. Of these, the approach of Coad and Yourdon [2] is notable for its simplicity and for the unity of its object oriented model[1]. Most of the other approaches listed use more than one model to analyse a system and are far closer in flavour to traditional structured analysis, which is both their strength and their failing.

Our approach to requirements specification is to use object oriented analysis[2] to provide the basic structure, and to use an object oriented extension of Z to provide the detail. Consequently, the diagrams of object oriented analysis provide the primary specification, but, where greater precision is needed, the power of Z is available.

In section 7.2 we look at the advantages and shortcomings of traditional structured analysis. We then see how the shortcomings have been partly answered in object oriented analysis.

In section 7.3 we see what mathematically based languages such as Z can offer that is different from structured or object oriented analysis. We then discuss object oriented extensions of Z which will enable us to add greater precision to our diagrammatic specifications.

Section 7.4 surveys a selection of object oriented analysis techniques. Section 7.5 explains how object oriented Z and analysis can be used together, and section 7.6 indicates some ways in which the formal parts of such a mixed specification can be used to aid validation.

In section 7.7 we apply our ideas to specifying a basic telephone service, while section 7.8 summarizes related work on structured analysis. In section 7.9 we draw conclusions and in section 7.10 suggest future work to develop and try out the technique.

7.2   FROM STRUCTURED TO OBJECT ORIENTED ANALYSIS

Structured analysis is a term we use to cover an approach to software development which:

- systematically decomposes a system into component parts, treating data and processes separately;

---

[1] This is in no way related to Yourdon structured analysis apart from the involvement of Edward Yourdon in both pieces of work.

[2] We have used that of Coad and Yourdon [2] although other approaches could be used.

- describes the resulting components using some prescribed notation (often diagrammatic);

- prescribes the various activities which must be performed as part of the analysis.

CASE tools are provided for many such structured analysis methods and it is usual for companies to market proprietary methods in tandem with appropriate CASE tools. Structured analysis methods are usually linked to design methods and in some cases form part of a total development life cycle methodology.

In this chapter we will be concerned only with the analysis phase, although it is not always easy to distinguish between analysis and design.

The archetypal structured analysis method advocates a number of complementary ways of analysing a customer's problem. The results of each analysis will be recorded with the aid of diagrams in a prescribed way. Various checks and cross checks will then be carried out on the digrams to look for inconsistencies and it may be that as a result changes are made. Finally the results of the analysis will provide input to the design stage.

Two popular methods commonly in use in the UK are the Structured System Analysis and Design Method (SSADM) [9], which is required for many government software contracts, and the structured analysis of Yourdon [10].

Among the advantages claimed for such methods are that:

- they impose a discipline on software development and if followed conscientiously can produce adequate documentation;

- the diagrams produced can be understood intuitively by non-experts, especially the customer;

- they help to turn software development into a skill that can be taught rather than an activity that depends on inspiration.

Among the weaknesses sometimes observed are that:

- they provide separate models of data and processes which may not be easy to relate;

- each modification of a specification may involve changes to many different parts of the documentation;

- the monolithic nature of the specification makes the development life cycle rigid;

- the diagrammatic notations used, though intuitively appealing, are never rigorously defined and usually depend on informal descriptions;

- the size of specifications can be very large.

Object oriented analysis methods are similar in some respects to structured methods in that:

- they analyse the system by breaking it down into components;

- they use diagrams intended to be understood by the non-expert.

Where they differ is that they are based on objects and classes, as defined in Chapter 1. An object is usually described by a collection of data called the attributes. Each class has a collection of operations which can change or read this data. A key idea is that the attributes of an object can only be changed or read by the specified operations. The consequences of this are that:

- once a class has been specified at the analysis stage, it can be treated independently all the way through to code, hence providing greater flexibility in the development life cycle;

- because data and the processes affecting that data are treated together, the transition to design and then to code is more natural;

- the implementation of a class can be changed without affecting the behaviour of the system, provided that the behaviour of an object of that class does not change as viewed through the operations.

Object orientation also has powerful mechanisms for building new classes from old, in particular inheritance and the part-whole relationship, both described in Chapter 1.

As a result of such mechanisms, many people report that specifications are shorter and simpler. Moreover the added structure carries with it greater understanding of the system being specified.

It is commonly assumed that when object oriented analysis is used the final target programming language must also be object oriented. But this need not be so and even when using a conventional language, advantages can be got by using an object oriented approach (see Chapter 5). In conclusion, object oriented analysis offers the advantages of structured analysis but, by using objects, offers:

- shorter and simpler specifications;

- specifications which are more easily modified;

- a development life cycle which is more flexible;

- specifications which map cleanly and naturally on to the design and final code.

Many people have published books on object oriented analysis and design in recent years. It is worth mentioning here that there is no clear line between what is analysis and what is design. Booch [5], for example, calls both processes together design. Many writers make the point that object orientation blurs the distinction and certainly makes it possible to proceed well into the implementation stage on one class while another class may still be in the analysis stage. In this chapter we only deal with the analysis stage.

7.3   Z AND OBJECT ORIENTED Z

Z is a specification language based on set-theory [1]. It is very simple in principle and yet very powerful in practice. Although it is not essential to have a deep knowledge of Z to read this chapter, some understanding of the basic concepts of the language will be helpful.

The system to be modelled is treated as a set of states which it can assume together with a set of operations, each of which may:

- change the state of the system;

- receive data from the external world (inputs);

- send data to the external world (outputs).

A very useful mechanism in Z is that of the **schema**. A schema consists of a box with one or two compartments. The first compartment lists the variables to be used, and for each variable names a set to which that variable belongs. The second compartment, which is optional, describes symbolically a relationship between those variables. A simple example is given in Fig. 7.1. This schema tells us that we are going to use the variable x, that x belongs to the set, N, where N represents the set of positive whole numbers and the colon means 'belongs to', and that x satisfies the condition $x \leq 5$. In general there will be more than one variable in the top compartment and the bottom compartment may be empty.

```
┌─ Example ──────────────────────────────┐
│  x : ℕ                                  │
│ ┌──────────────────┐                    │
│  x ≤ 5                                   │
└─────────────────────────────────────────┘
```

Fig. 7.1   Simple schema.

A state of the system is usually represented by assigning values to variables. An operation can then be specified by a schema which relates the values of the variables before and after the operation. Suppose that the states are all represented by whole numbers. Then the operation in Fig. 7.2 adds 1 to the state. It is a convention that variables representing states after the operation carry dashes. There are many notational conventions which we have not mentioned which add considerably to the power of the language. One convention, which we will need, is that inputs carry a question mark after them and outputs carry an exclamation mark.

```
┌─ Increment ────────────────────────────┐
│  x : ℕ                                  │
│  x' : ℕ                                 │
│ ┌──────────────────┐                    │
│  x' = x + 1                              │
└─────────────────────────────────────────┘
```

Fig. 7.2   Simple operation.

There are at least two reasons for wanting to write object oriented specifications. Firstly they provide a better start if other parts of the development are to be object oriented. Secondly, and maybe more importantly [11], they provide an excellent way of structuring large and complex systems so as to make them more manageable. BT has many such systems and hence object oriented specification is particularly appropriate to its needs.

Powerful though Z is, it is not totally suitable for writing object oriented specifications. In particular objects and classes are not easy to represent satisfactorily. In order to be able to write natural object oriented specifications, a number of people have suggested extensions to Z[11].

Here we refer to two particular extensions, Object Z[12] and ZEST[13, 14], the latter being in part a development of earlier work by Cusack on object orientation [15]. Both of these languages specify each class as though it were a Z specification in its own right. Each language is still evolving and has many powerful constructs. For the purposes of this chapter, however, we have adopted a simplified notation which we will call object oriented Z. Each class has a set of states specified by a state schema, an initialization schema and a set of operations which act on that state, each operation being specified by an operation schema. All the schemas associated with a given class are collected together in a single box called the class schema as in Fig. 7.3.

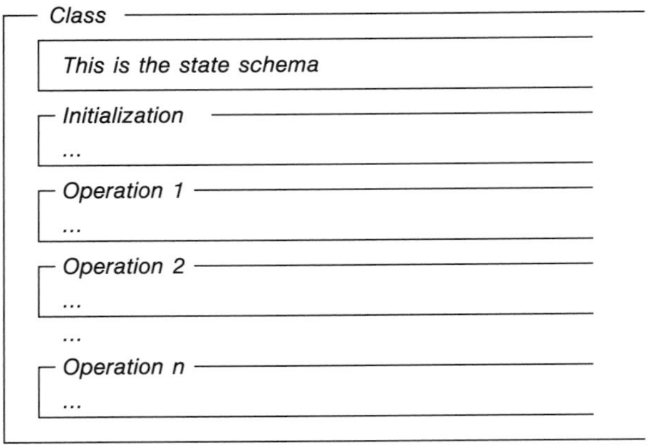

Fig. 7.3   Class schema in object oriented Z.

Later on, when discussing object oriented analysis, we will need to talk about message connections.

We can think of a message as follows — a request is sent to another object asking for a particular operation to be performed on that object; it may well be that as a result of this, information is passed back. There are two ways in which messages might be passed in object oriented Z.

When defining an operation in one class it is sometimes necessary to refer to operations in other classes. Suppose that *a* is an object of a class *A* and *Op* is an operation of *A*. If *a* is referred to in another class, then *a.Op* can be used to refer to the operation schema *Op* when acting on *a*. This kind of message passing can be likened to a functional call.

Another kind of message passing can be represented by inputs and outputs. ZEST does not at present provide this kind of message passing. Object Z, though it does provide message passing of this kind, does it in a rather different way than that used here. Our approach has been adopted for the sake of simplicity. If we want to pass a piece of data between two objects, then we represent it by the same variable in both objects but terminated with a ! to indicate the source, and a ? to represent the destination.

7.4   A SURVEY OF SOME OBJECT ORIENTED ANALYSIS TECHNIQUES

All the approaches to object oriented analysis that we have looked at give very similar information, but structured and recorded in very different ways. It is not our purpose here to examine in detail the fine distinctions between one approach and another or award points. This has been done by other writers. Our objective is to survey a selection of methods briefly to see how they record and structure the information we need when writing our object oriented Z specification.

We look at four different approaches — Coad and Yourdon [2], Booch [5], Martin and Odell [6] and Rumbaugh [3]. We have surveyed the approach of Shlaer and Mellor [4], but since it is very close to that of Rumbaugh we have not discussed it further.

7.4.1 Coad and Yourdon

The Coad and Yourdon approach is unusual because it presents one integrated model. Although the model has five layers, the notation for each layer is strikingly simple and the diagrams for each layer fit together into one diagram, which gives most essential information which can in our opinion be usefully recorded in such form. We will describe each layer and use this as an opportunity to introduce the concepts of object orientation. We should think of each successive layer as providing yet more information in our model.

At the top is the subject layer. The concept of a subject is not part of object orientation as such. It is a grouping of related classes. Hence classes which are related in the structure should properly be in the same subject. The purpose of subjects is to make the model easier to grasp and to aid organization for larger projects. Subjects are indicated by enclosing the related classes in a box. Although this layer logically comes first, it cannot really be built until the classes and their structures have been determined. This is done in the next two layers.

The class layer[1] consists of all the classes of objects which make up the system. Determining such classes is the first and possibly most important part of the analysis. There are many different ways of defining a class. Perhaps the most helpful way is to think of it as a generic description of a number of objects. A class, for example, might be *BTCustomer* and would be represented as in Fig. 7.4. An object of the class would be an actual BT customer, such as the author of this chapter.

The structure layer records two types of relationship between classes namely generalization-specialization (gen-spec) and whole-part.

A class *A* is a specialization of class *B* if each object which fits the description of *A* also fits the description of *B*. The class *BTCustomer* has specializations *PrivateCustomer* and *BusinessCustomer* and can be represented as in Fig. 7.5. Notice that in Fig. 7.5 the representation of *BTCustomer* is

---

[1] Coad and Yourdon refer to this as the Class and Object layer, reserving the term Class for what we call an abstract class.

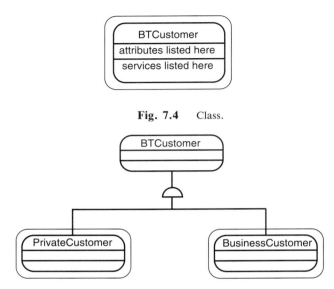

Fig. 7.4   Class.

Fig. 7.5   Generalization-specialization.

different from that in Fig. 7.4. This is because every BT customer is either a private customer or a business customer. Such a class is called abstract.

An object of a class *A* is a part of an object of a class *B* if each object of *A* uses an object of *B* in its definition. Two telephone *Users* are needed in order to define a telephone *Call*, the users being the person who calls (caller) and the person who is called (callee). Each *User* is a part of the *Call*. The structure is called a whole-part structure and is represented as shown in Fig. 7.6, referred to in Chapter 1 as the part-whole relationship.

The attribute layer defines the attributes of each class. An attribute can be thought of as what an object of that class 'needs to know about itself'. Attributes are listed in the class box as shown in Fig. 7.4. In some cases an object may need to know about other objects. This information is indicated by 'instance connections'. We can think of instance connections as telling us that a class has an object of another class as an attribute. An instance connection is indicated by a straight line joining the two classes.

The last layer, the service layer, describes the behaviour of the objects. Key concepts in object orientation are encapsulation and information hiding. By encapsulation we mean that the information about an object and the operations that can change or access that information are packaged together. Information hiding means that the values of the attributes can only be changed

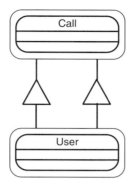

Fig. 7.6   Whole-part representation.

or accessed by those operations. Some people say that they form an interface with the object. Coad and Yourdon call these operations services. You could say that a service is 'what an object can do'. A system in each object that only changed or accessed its own information would be very unhelpful since the information would not then be visible outside the object. Hence most objects will need to use the services of other objects. This can be done using 'message connection' mentioned in section 7.3. Such a message connection is represented by a single arrow.

The specification of services is achieved by a mixture of textual descriptions and conventional diagramming, for example state transition diagrams. But it seems perfectly possible to adopt other techniques for this purpose without invalidating the rest of the approach. It is precisely in the definition of services that object oriented Z can add most value to object oriented analysis.

7.4.2   Booch

Booch [5] deals with object oriented software development all the way to implementation although he concentrates on analysis and design. It is not always easy to separate the two, but that part of the approach that principally addresses analysis is the construction of the class diagram and the object diagram. The class diagram describes the relationships between classes. In comparison with Coad and Yourdon's two structures, Booch has eight different types of arrow. Here as elsewhere Booch gains expressiveness at the expense of simplicity. In addition there are class and class-utility templates which reference operation templates, an operation being the same as a Coad and Yourdon service. Attributes are documented in the templates. The functionality of each operation is described by state transition diagrams, fragments of some 'program description language', which could be a programming language and object diagrams. The object diagram indicates the passing of messages between objects. There are five different types of arrow and two different types of template.

Booch's notation is very sophisticated and very detailed. In particular it provides good facilities for real time applications. Object oriented Z could be used to replace much of the templates, although there are details found in the templates which possibly relate to a design activity and would not normally be recorded in a requirements capture activity. Object oriented Z could also be used to replace the code fragments and state transition diagrams.

7.4.3   Martin and Odell

Martin is well known for his work on information engineering [16-18] which is characterised by its strong emphasis on business analysis. He and Odell [7] attempt to recast information engineering into an object oriented framework. Their approach is clearly divided into analysis and design. Like Booch's approach it has essentially two models — the object structure analysis (OSA) and the object behaviour analysis (OBA).

The OSA identifies classes together with their relationships. Such relationships include the equivalent of gen-spec and whole-part. They also include relationships corresponding to Coad and Yourdon's instance connections.

The OBA identifies the operations which allow the states of an object to be changed or accessed.

The approach of Martin and Odell to object orientation is markedly different from that adopted in this chapter. Firstly it leans heavily on entity-relationship modelling, although not as strongly as the approach of Shlaer and Mellor [4], and, secondly, an object can change its class. Certainly this would create serious difficulties when trying to treat a class formally.

The approach of Martin and Odell claims to offer an easy migration from information engineering to some form of object orientation. It stresses heavily the use of diagrams and attempts to record as much information as possible in diagrammatic form.

To use object oriented Z with Martin and Odell would not be easy, principally because the information relating to a class is spread between many different diagrams. It is probably at the operation level again that the most value can be added.

7.4.4   Rumbaugh

Rumbaugh and his associates [3] carry out their object oriented analysis very much along the lines of traditional structured analysis but in such a way that the results of the analysis can be related to an object oriented outlook. There are three models — the object model, the dynamic model and the functional model. The object model corresponds to the entity relationship model, the dynamic model describes the possible states of an object and how they can change, and the functional model describes the data-flows within the system.

Rumbaugh et al accept the similarity to traditional structured analysis but claim that the difference is that their emphasis is on object modelling whereas traditional structured analysis emphasizes process modelling.

The best way of using object oriented Z here would be to choose classes which were felt to be problematical and build complete specifications of them. From the object model would come information about attributes, gen-spec and whole-part information. From the functional model would come the states of an object and hence the state variables. The services could be identified from the functional model. The dynamic and functional model go beyond what is normally modelled in object oriented Z, but, if it were required to model all the information, it could be done.

7.5   HOW OBJECT ORIENTED Z AND THE OBJECT ORIENTED ANALYSIS OF COAD AND YOURDON ARE RELATED

In this section we describe how the notation of Coad and Yourdon's object oriented analysis (CY) and object oriented Z can be related. An understanding of this relationship will enable us either to use CY as a first stage when writing an object oriented Z specification or to add fragments of object oriented Z to an existing CY analysis. It should, however, be realized that the formal specification in object oriented Z cannot be derived mechanically from the CY analysis, since many parts of the CY analysis are given informally.

In the next section we will give a partial CY analysis of the basic telephone call together with a full object oriented Z specification of one of the classes. It should be noted that this is intended to model what happens as seen by the end users and may not reflect how a telephone engineer would view the system.

In CY there are five activities corresponding to each of the five layers. The first three activities, namely finding class-and-objects, identifying structures and identifying subjects, need to be completed before starting work on the formal specification. The last two, defining the attributes[1] and services, are done while writing the formal specifications.

A class-and-object or class will be described in object oriented Z by a class schema, which will be built using information from the structure layer and the attribute and service layers. This has to be done iteratively as parts of the formal specification of one class will often be needed for specifying another class.

There are two types of structure identified in CY — gen-spec structures and whole-part structures. For each of these structures we see how they are reflected in the class schema.

For gen-spec structures we must first identify the most general classes, i.e. those that are not specialization classes within some gen-spec structure. These should be specified before their specializations and will be done using schema inclusion. Suppose that a generalization and specialization are indicated in Fig. 7.7. Then if G is the class schema of the generalization and S the class schema of the specialization, then this would be specified as in Fig. 7.8.

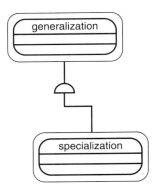

Fig. 7.7   Specialization-generalization.

```
┌─ S ─────────────────────────────────────────────
│  G
│  Extra attributes from the specialization
│  Extra service specifications from the specialization
└──────────────────────────────────────────────────
```

Fig. 7.8   Specialization-generalization in object oriented Z.

[1] ZEST [13, 14] makes a distinction between those attributes which cannot be changed by the operations of a class, and those attributes, called state variables, which can be. Such a distinction does not exist in Object Z. To make the presentation easier to understand we do not make this distinction, although it does not present any insurmountable technical problems.

In the case of whole-part structures, we must identify those classes which cannot be built from other specified classes. Suppose that *Part*1 and *Part*2 are parts of *Whole* and each object of *Whole* needs two objects of *Part*1 and three objects of *Part*2 to do its job. This is described by the Coad and Yourdon analysis shown in Fig. 7.9. The class schema *W* of *Whole* will then be as in Fig. 7.10, where *P*1 is the class schema of *Part*1 and *P*2 is the class schema of *Part*2.

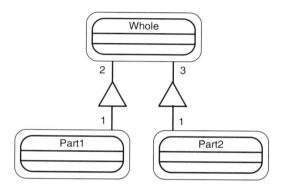

Fig. 7.9   Whole-part structure.

```
┌─ W ──────────────────────────────────────────────────┐
│  ┌──────────────────────────────────────────────────┐ │
│  │                                                    │ │
│  ├──────────────────────────────────────────────────┤ │
│  │  set_of_P1 : ℙ P1                                 │ │
│  │  set_of_P2 : ℙ P2                                 │ │
│  │  ────────────────────────────────────────────    │ │
│  │  #set_of_P1 = 2                                   │ │
│  │  #set_of_P1 = 3                                   │ │
│  │  Other constraints on W                           │ │
│  └──────────────────────────────────────────────────┘ │
│  Services of Whole                                     │
└───────────────────────────────────────────────────────┘
```

Fig. 7.10   Whole-part structure in object oriented Z.

Here $P\,S$ represents the set of all subsets of S, and $\#S$ represents the number of elements in S. This means that *set_of_P*1 is a pair of objects from *P*1 and *set_of_P*2 is a set of three objects from *P*2.

A subject in CY is identified by finding the uppermost class in a structure hierarchy. Hence a subject could form a chapter or section, say, of the formal specification.

It is the attribute and service layers which have the most information to give for the formal specification.

When specifying attributes it is necessary to:

- say to which sets attributes belong;

- give any constraints on the attributes.

Whether or not this information can be found in the object oriented analysis will depend on how thorough it is. The sets to which the attributes may belong will be described in terms of other sets defined elsewhere and possibly including given sets and other classes. If it is not clear how to specify an attribute formally, then there is information missing in the analysis.

Instance connections form part of the attribute layer because they help to answer the question 'what does the object know about itself?' The instance connection shown in Fig. 7.11 indicates that A may need an object of B in order to function correctly. Another way of looking at this is to say that A may have an object of B as an attribute. Hence the state schema of A would be of the form given in Fig. 7.12. In the case of one-to-many connections the attribute would be a set of objects and if there were more detailed information on how many objects of A correspond to how many objects of B then this would be included in the Z specification as a constraint. In the case of many-to-many instance connections we would need to introduce an associative class, as described for example by Shlaer and Mellor [4] and as in traditional data analysis. Such an associative class would have objects of A and B as attributes. It is important when introducing such classes to understand to what they correspond in the real world and not just to introduce them as a technical convenience.

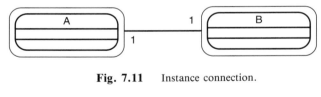

Fig. 7.11   Instance connection.

$b : B$
Other attributes

Fig. 7.12   Instance connection in object oriented Z.

Finally we deal with the service layer. Let us suppose that we now have an object oriented Z class description which has attributes. We now need to fill in the operations. From the analysis we should have:

- the names of services;

- message connections from the service layer as marked on the diagram;

- service specifications in the class template.

If in CY a message connection joins class *A* and class *B*, then some service of *A* must be using a service of *B*. There are two ways in which such a message connection can be modelled.

The first way is suitable when an object *b* of *B* is referred to in some way within *A*, e.g. as an attribute. Suppose that *Op* is the operation schema for the service to be accessed in *B*. Then *b.Op* can be used in *A* to describe the required service acting on *b*.

The second way models the passing of data between two objects more directly. Suppose that an object of *A* passes data $x_1, x_2 \ldots x_m$ to an object of *B* which responds by sending back data $y_1, y_2 \ldots y_m$. Then the data passed from *A* to *B* will be denoted as outputs $x_1?, x_2? \ldots x_m?$ from an appropriate operation of *A* and as inputs $x_1!, x_2! \ldots x_m!$ to an appropriate operation of *B* with the convention that, when a variable name occurs with ? in the class description of *A* and the same variable name occurs with ! in the class description of *B*, then they have the same value. The data being passed back from *B* to *A* will be represented in a similar way. It is important to realize that such input and output variables are visible throughout the whole specification and are uniquely associated with a particular message connection.

7.6   FEEDBACK FROM THE FORMAL SPECIFICATIONS

Formal specification can be used to help in validating the requirements specification. The following suggestions by no means cover all possibilities but give some idea of what can be done.

An ordinary Z specification can be type-checked. In Z everything has a unique type. If that type cannot be determined or is ambiguous then there is something wrong with the specification. Type-checking for Z can be carried out automatically. Although type-checking for whole object oriented Z specifications is not yet available, it is possible to carry out some degree of type-checking on the operations within individual classes.

Every operation in Z has a pre-condition, i.e. a formal condition which a state must satisfy in order for the operation to be applied to it. In practice it can usually be simplified to give precise conditions under which the operation will succeed. By interpreting such pre-conditions and feeding them back to the customer, we can check whether all possible situations have been covered.

Z is not a programming language and not directly executable [19]. Its purpose is to allow one to describe the behaviour required of systems when regarded as black boxes. The process of obtaining a working program from a Z specification requires design decisions on the part of the developer [20].

Nevertheless it seems to the author that there is scope for building useful tools for helping in the validation of Z specifications against requirements [21, 22].

Finally the very fact that one can build a formal model of a system gives us greater confidence in the internal consistency of our requirements.

7.7 APPLICATION TO THE BASIC TELEPHONE CALL

In this section we take as an example the basic telephone call. We do this to provide an illustration of how our ideas might be applied. Neither the analysis nor the object oriented Z specification are intended to be complete. Not even in a full-scale analysis might an object oriented Z specification be necessary for all parts of the model. How much of the model to specify formally would depend on the nature of each class and its relative importance to the success of the project. We have tried to keep the notation as simple as possible, but those special symbols occurring in this section and not explained in the text are given in Appendix B. For reference we include a full object oriented Z specification as Appendix A, but this is intended for the reader with a fuller knowledge of Z.

The analysis in this example is carried out from the point of view of the users. We provide a Coad and Yourdon diagram and descriptions in object oriented Z of selected classes. We first describe in words the system to be modelled. Then we explain the CY analysis drawing and finally we provide formal specifications of the classes Network and Connection.

A caller lifts the handset and dials the number of the callee, the person he wishes to call. As soon as he has dialled a number which is recognized as obtainable, then, if the callee's telephone is not busy, it rings and the caller hears a ringing tone. If, the callee's telephone is busy then the caller hears a busy-tone. If, when dialling, the caller dials an unobtainable number, then he will hear an NU-tone (number unobtainable).

Let us suppose now that the caller has dialled an obtainable number and that the callee's telephone is not busy. When the callee picks up the handset, the caller and callee are now able to talk to each other, i.e. their telephones are connected.

There are many ways of modelling the above system using object oriented concepts. We have chosen to view the system as a number of Devices, the Network and a Directory. A device in our model can be a Subscriber's telephone or a tone of some sort (ring-tone, dial-tone, busy-tone or NU-tone) or a telephone ringing. A Connection can be either between two subscribers or between a subscriber and another device. A connection between a subscriber and the dial-tone, for example, means that the subscriber is hearing the dial-tone. That part of the system which establishes connections between devices will be called the Network. There is only one object in the class Network. It will keep a record of all current connections. Finally there will be a Directory which knows which subscribers have which telephone numbers.

We now look at the features of the analysis drawing. Each Connection is seen as a whole which has two parts, each of which is a Device. A Device can be a subscriber, a phone_ringing, a dial_tone, a ring_tone, a busy_tone or an NU_tone. Hence Device is an abstract class. There is an instance connection between Network and Connection, since one of the attributes of Network is a set of connections. There is also an instance connection between each instance of Network and each instance of Directory, a Directory keeping a table of telephone numbers and their related subscribers. The other attributes of the Network are a number_buffer in which telephone numbers are accumulated as they are dialled by subscribers, and the set of all subscribers who are allowed to telephone out.

There are a number of message connections between instances of Subscriber and of Network but only one connection has been recorded in the diagram. Subscriber has three services, off_hook, send_digit and on_hook. Network has three corresponding services — receive_off_hook_signal, receive_digit and receive_on_hook_signal. These will use other services — connect, disconnect, add and initialize_number_buffer; connect connects two given Devices; disconnect disconnects two given Devices; add adds a dialled digit to a given entry in the number_buffer; initialize initializes a given entry in the number_buffer. There are also message connections between instances of Network and of Directory—receive_digit uses the services lookup and check of Directory; lookup finds the subscriber with a given telephone number; check determines whether or not a given sequence of digits is part of a valid number and, if part of a valid number, whether the number is complete. The analysis drawing of the system is given in Fig. 7.13.

We now sketch the object oriented Z specification of this system. Although a complete specification has been written we will only give parts of it to illustrate how the specification has been constructed.

It is important when writing a formal specification to be able to identify the various objects in the different class. For this reason we need a given set:

[ID]

of identifiers for subscribers. As noted above, *Device* is a class and every object of *Device* is a *Subscriber* or *phone_ringing* or some kind of tone. This is represented in Z by using a so-called free type. Free types enable one to

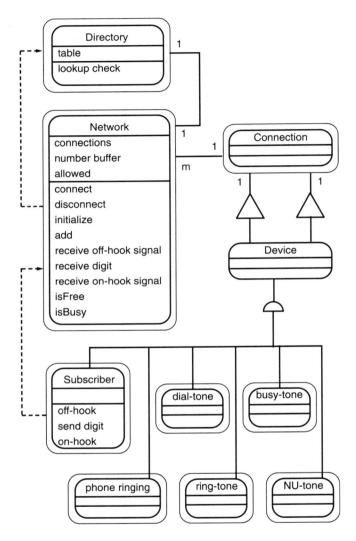

Fig. 7.13 Basic telephone call.

list a number of mutually exclusive alternatives. Each of the alternatives can be parameterized by some other set. In the case of *Device*, *subscriber* is parameterized by *ID*:

Device__id :: = *subscriber* ⟪*ID*⟫
| *phone__ringing*
| *dial__tone*
| *ring__tone*
| *busy__tone*
| *NU__tone*

The other given sets needed are:

[*Message*]

to represent on-hook and off-hook signals, which we will not need in our partial specification given here, and:

[*DIGIT*]

to represent telephone digits. We also need the free type, *Status*:

Status :: = COMPLETE | INCOMPLETE | INVALID

which is needed to describe the outcome of checking a sequence of digits to find out if it forms part of a recognized telephone number.

Firstly, we specify the class *Subscriber*:

```
┌─ Subscriber ─────────────────────────────────
│  ┌──────────────────────────────────────────
│  │ subscriber__id : ID
│  ├──────────────────────────────────────────
│  ┌─ off__hook ──────────────────────────────
│  │ ...
│  └──────────────────────────────────────────
│  ┌─ send__digit ────────────────────────────
│  │ digit! : DIGIT
│  │ A__subscriber__id! : ID
│  ├──────────────────────────────────────────
│  │ A__subscriber__id! = subscriber__id
│  └──────────────────────────────────────────
│  ┌─ on__hook ───────────────────────────────
│  │ ...
│  └──────────────────────────────────────────
└───────────────────────────────────────────────
```

The state of a subscriber is given by the *subscriber__id*. The operation *send__digit* sends out a single digit to the *Network* together with its own identifier. This digit will be received by the service *receive__digit* in *Network*, which is specified as follows:

```
┌─ Network ──────────────────────────────────────────────
│
│  ┌─────────────────────────────────────────────────────
│  │ connections : P Connection
│  │ number__buffer : ID → seq(DIGIT)
│  │ allowed : P ID
│  │ directory : Directory
│  ├─────────────────────────────────────────────────────
│  │ ┌─ connect ──────────────────────────────────────────
│  │ │ Δ(connections)
│  │ │ caller?, callee? : Device__id
│  │ ├─────────────────────────────────────────────────────
│  │ │ (∃connection : Connection •
│  │ │   (connection.caller = caller?∧
│  │ │    connection.callee = callee?∧
│  │ │    connections' = connections ∪ {connection} ))
│  │ └─────────────────────────────────────────────────────
│  │   ...
│  │ ┌─ add ──────────────────────────────────────────────
│  │ │   ...
│  │ └─────────────────────────────────────────────────────
│  │ ┌─ receive__digit ───────────────────────────────────
│  │ │ Δ(connections)
│  │ │ A__subscriber__id?, B__subscriber__id? : ID
│  │ │ digit? : DIGIT
│  │ │ number__status? : Status
│  │ │ number! : seq(DIGIT)
│  │ │ subscriber__id? : ID
│  │ ├─────────────────────────────────────────────────────
│  │ │ add
│  │ │ ((number__status? = COMPLETE ∧
│  │ │ ((isFree ∧
│  │ │ connect[subscriber A__subscriber__id? / caller?,
│  │ │  ring__tone / callee?] ∧
│  │ │ connect[subscriber B__subscriber__id? / caller?,
│  │ │  phone__ringing / callee?])
│  │ │     ∨
│  │ │ (isBusy ∧
│  │ │ connect[subscriber A__subscriber__id? / caller?,
│  │ │  busy__tone / callee?])))∨
│  │ │ (number__status? = INCOMPLETE ∧
│  │ │ connections' = connections)
│  │ │     ∨
│  │ │ number__status? = INVALID ∧
│  │ │ connect[subscriber A__subscriber__id? / caller?,
│  │ │  NU__tone / callee?])
│  │ └─────────────────────────────────────────────────────
│  │   ...
│  │ ┌─ IsFree ───────────────────────────────────────────
│  │ │   ...
│  │ └─────────────────────────────────────────────────────
│  └─────────────────────────────────────────────────────
```

receive__digit takes a digit as input from a subscriber with identifier *A__subscriber__id?*. This digit is added to the appropriate part of the number buffer. A message is then sent to the *Directory* to find out the status of the sequence of digits received so far. If the sequence is a complete telephone number, then it returns the identifier of the subscriber corresponding to that number, which is given by *B__subscriber__id?*. There are two local services *isFree* and *isBusy* which check whether this subscriber is free or not. If the subscriber is free then the caller is connected to the *ring__tone* and the subscriber being called is connected to *phone__ringing*. These connections are made with *connect* and are modelled as a set of instances of the class, *Connection*, which is specified by:

```
┌─ Connection ──────────────────────────────
│ ┌────────────────────────────────────────
│ │ caller : Device__id
│ │ callee : Device__id
│ │
```

Finally we look at the class, *Directory*, specified as follows:

```
┌─ Directory ───────────────────────────────
│ ┌────────────────────────────────────────
│ │ table : seq(DIGIT) → ID
│ │ ┌─ lookup ───────────────────────────
│ │ │ number? : seq(DIGIT)
│ │ │ B__subscriber__id! : ID
│ │ │
│ │ │ B__subscriber__id! = table number?
│ │ ┌─ check ────────────────────────────
│ │ │ number? : seq(DIGIT)
│ │ │ number__status! : Status
│ │ │
│ │ │ (number? ∉ ∪(initial__segments((dom table))) ∧
│ │ │ number__status! = INVALID)
│ │ │ ∨
│ │ │ (number? ∈ (dom table) ∧
│ │ │ number__status! = COMPLETE)
│ │ │ ∨
│ │ │ number__status! = INCOMPLETE
```

lookup receives a telephone number as input and returns the identifier for the corresponding subscriber. *check* receives a sequence of digits as input and decides whether it is complete, incomplete or invalid. It is invalid if it does not occur as an initial segment of a recognized telephone number. The global function *initial__segments* is assumed to have been defined elsewhere, but presents no technical difficulties.

7.8   OTHER WORK ON INTEGRATION

There has been considerable interest in recent years in trying to integrate structured analysis techniques and Z. Semmens and Allen [23] describe a development method which combines the structured analysis of Yourdon and ordinary Z. As part of the SAZ project at York University, Polack, Whiston and Hitchcock [24] have developed a method for writing Z specifications using SSADM version 4. There has also been work carried out by Merad [25] to integrate Z and Shlaer-Mellor object oriented analysis.

Giovanni and Iachini [26] propose a technique for using Z to provide more formality to HOOD. HOOD is a development method used by the European Space Agency. It is object based rather than object oriented, meaning that it does not have classes or relationships between classes.

Tse [27] tries to give more formality to the notations of structured analysis by building a common algebraic framework. It is conceivable that a similar exercise might be carried out for object oriented analysis.

7.9   CONCLUSIONS

The aim of this chapter has been to show how a comparatively compact and yet precise notation based on set theory can be used to give greater value to object oriented analysis. Both diagrammatic and symbolic notations have their value, the former being more intuitively appealing and more accessible provided that they are not too complicated, the latter being more compact, easier to manipulate mathematically and often more expressive.

Object oriented analysis can also be used to provide a systematic approach to writing object oriented Z specifications.

Whether we write the whole specification in object oriented Z or just selected classes, any such fragments of formal specification can be used to aid validation by applying type-checking, examining pre-conditions and providing greater confidence in internal consistency. Commercial tools already exist for editing and type-checking Z specifications. Although no commercial tools exist at present for animating Z specifications, there has been work in this area and it is technically feasible to do this as an aid for validation for a useful subset of the language.

7.10   FUTURE DEVELOPMENTS

The techniques described in this chapter need to be applied to a wide range of projects if a better understanding is to be gained of how they apply in practice and to what extent they need to be modified or extended. Object oriented development is being tried in a number of areas and there is considerable scope for using these ideas.

More work needs to be done on applying object oriented Z to the problems of validation. We know of no work on executing object oriented Z specifications and this is an area which deserves attention. There has, however, been some interesting work by Rafsanjani and Colwill [28], in which a structural mapping has been constructed from Object Z to C + +.

If our techniques are to be used on large projects, there needs to be more work done on the provision of CASE tools and such work would need to be supported by a better understanding of the theoretical relationship between object oriented analysis and object oriented Z.

Finally there needs to be an examination of ways of extending object oriented Z to express requirements which are at present outside its scope.

APPENDIX A

Full object oriented specification of basic telephone call

In practice such a specification should include explanatory text, but in this case this would have resulted in much duplication and would have considerably lengthened this appendix.

[*Message,ID,DIGIT*]

$$Device\_id:: = subscriber\langle\langle ID \rangle\rangle$$
$$\mid phone\_ringing$$
$$\mid dial\_tone$$
$$\mid ring\_tone$$
$$\mid busy\_tone$$
$$\mid NU\_tone$$

$$Status:: = COMPLETE \mid INCOMPLETE \mid INVALID$$

$$\begin{array}{l} \underline{[X]} \\ \hline initial\_segments : seq(X) \rightarrow \mathbb{P}(seq(X)) \\ \hline \forall s : seq(X) \bullet initial\_segments(s) = \{s1 : seq(X) \mid s1 \subset s\} \end{array}$$

```
┌─ Subscriber ──────────────────────────────────────┐
│  ┌─────────────────────────────────────────────┐  │
│  │  subscriber__id : ID                         │  │
│  ├─────────────────────────────────────────────┤  │
│  │  ┌─ off__hook ───────────────────────────┐   │  │
│  │  │  off__hook__signal! : Message          │   │  │
│  │  │  A__subscriber__id! : ID               │   │  │
│  │  ├────────────────────────────────────────┤   │  │
│  │  │  A__subscriber__id! = subscriber__id   │   │  │
│  │  └────────────────────────────────────────┘   │  │
│  │  ┌─ send__digit ─────────────────────────┐    │  │
│  │  │  digit! : DIGIT                        │    │  │
│  │  │  A__subscriber__id : ID               │    │  │
│  │  ├───────────────────────────────────────┤    │  │
│  │  │  A__subscriber__id! = subscriber__id  │    │  │
│  │  └───────────────────────────────────────┘    │  │
│  │  ┌─ on__hook ────────────────────────────┐    │  │
│  │  │  on__hook__signal! : Message          │    │  │
│  │  │  subscriber__id! : ID                 │    │  │
│  │  ├───────────────────────────────────────┤    │  │
│  │  │  subscriber__id! = subscriber__id     │    │  │
│  │  └───────────────────────────────────────┘    │  │
│  └─────────────────────────────────────────────┘  │
└────────────────────────────────────────────────────┘
```

```
┌─ Connection ──────────────────────────────────────┐
│  ┌─────────────────────────────────────────────┐  │
│  │  caller : Device__id                         │  │
│  │  callee : Device__id                         │  │
│  └─────────────────────────────────────────────┘  │
└────────────────────────────────────────────────────┘
```

```
┌─ Directory ───────────────────────────────────────┐
│  ┌─────────────────────────────────────────────┐  │
│  │  table : seq(DIGIT) ⇸ ID                     │  │
│  ├─────────────────────────────────────────────┤  │
│  │  ┌─ lookup ──────────────────────────────┐   │  │
│  │  │  number? : seq(DIGIT)                  │   │  │
│  │  │  B__subscriber__id! : ID              │   │  │
│  │  ├───────────────────────────────────────┤   │  │
│  │  │  B__subscriber__id! = table number?   │   │  │
│  │  └───────────────────────────────────────┘   │  │
│  │  ┌─ check ───────────────────────────────┐   │  │
│  │  │  number? : seq(DIGIT)                  │   │  │
│  │  │  number__status! : Status             │   │  │
│  │  ├───────────────────────────────────────┤   │  │
│  │  │  (number? ∉ ⋃(initial__segments⟨(dom table)⟩)) ∧  │   │  │
│  │  │  number__status! = INVALID)           │   │  │
│  │  │  ∨                                     │   │  │
│  │  │  (number? ∈ (dom table) ∧             │   │  │
│  │  │  number__status! = COMPLETE)          │   │  │
│  │  │  ∨                                     │   │  │
│  │  │  number__status! = INCOMPLETE         │   │  │
│  │  └───────────────────────────────────────┘   │  │
│  └─────────────────────────────────────────────┘  │
└────────────────────────────────────────────────────┘
```

```
┌─ Network ──────────────────────────────────────────┐
│ connections : ℙ Connection                          │
│ number_buffer : ID  ⇸  seq(DIGIT)                   │
│ allowed : ℙ ID                                      │
│ directory : Directory                               │
│ ┌─ connect ────────────────────────────────────┐   │
│ │ Δ(connections)                                │   │
│ │ caller?, callee? : Device_id                  │   │
│ ├───────────────────────────────────────────────┤   │
│ │ (∃ connection : Connection•                   │   │
│ │   (connection.caller = caller? ∧              │   │
│ │   connection.callee = callee? ∧               │   │
│ │   (connections' = connections ∪ {connection})))│  │
│ └───────────────────────────────────────────────┘   │
│ ┌─ disconnect ─────────────────────────────────┐   │
│ │ Δ(connections)                                │   │
│ │ caller?, callee? : Device_id                  │   │
│ ├───────────────────────────────────────────────┤   │
│ │ ∃connection : Connection•                     │   │
│ │   (connection.caller = caller? ∧              │   │
│ │   connection.callee = callee? ∧               │   │
│ │   connections' = connections \ {connection})  │   │
│ │ number_buffer' = number_buffer                │   │
│ └───────────────────────────────────────────────┘   │
│ ┌─ initialize_number_buffer ───────────────────┐   │
│ │ Δ (number_buffer)                             │   │
│ │ subscriber_id? : ID                           │   │
│ ├───────────────────────────────────────────────┤   │
│ │ number_buffer' = number_buffer ⊕ {subscriber_id? ↦ ⟨⟩} │
│ └───────────────────────────────────────────────┘   │
│ ┌─ add ────────────────────────────────────────┐   │
│ │ Δ (number_buffer)                             │   │
│ │ digit? : DIGIT                                │   │
│ │ subscriber_id? : ID                           │   │
│ ├───────────────────────────────────────────────┤   │
│ │ number_buffer' =                              │   │
│ │   number_buffer ⊕                             │   │
│ │   {(subscriber_id? ↦                          │   │
│ │     (number_buffer subscriber_id?)⌢ ⟨digit?⟩)}│   │
│ └───────────────────────────────────────────────┘   │
│ ┌─ isFree ─────────────────────────────────────┐   │
│ │ Ξ_subscriber_id? : ID                         │   │
│ ├───────────────────────────────────────────────┤   │
│ │ ¬(∃connection : connections•                  │   │
│ │   ((subscriber Ξ_subscriber_id? = connection.callee)∨│
│ │   (subscriber Ξ_subscriber_id? = connection.caller)))│
│ └───────────────────────────────────────────────┘   │
│ ┌─ isBusy ─────────────────────────────────────┐   │
│ │ Ξ_subscriber_id? : ID                         │   │
│ ├───────────────────────────────────────────────┤   │
│ │ ∃connection : connections•                    │   │
│ │   (subscriber Ξ_subscriber_id? = connection.callee∨│
│ │   subscriber Ξ_subscriber_id? = connection.caller)│
│ └───────────────────────────────────────────────┘   │
└─────────────────────────────────────────────────────┘
```

```
┌─ receive__off__hook__signal ──────────────────
│ Δ(connections)
│ off__hook__signal? : Message
│ A__subscriber__id? : ID
├────────────────────────────────────────────────
│ A__subscriber__id?∈ allowed
│ ((∃connection : Connection•
│   (connection.caller = subscriber A__subscriber__id? ∧
│    connection.callee = phone__ringing ∧
│    connect[subscriber A__subscriber__id? / caller?,
│      directory.table (number__buffer A__subscriber__id?)
│       / callee?]))
│    ∨
│   (connect[subscriber A__subscriber__id? / caller?,
│    dial__tone / callee?] ∧
│    initialise__number__buffer))
└────────────────────────────────────────────────
```

```
┌─ receive__digit ──────────────────────────────
│ Δ(connections)
│ A__subscriber__id?, B__subscriber__id? : ID
│ digit? : DIGIT
│ number__status? : Status
│ number! : seq(DIGIT)
│ subscriber__id? : ID
├────────────────────────────────────────────────
│ add
│ ((number__status? = COMPLETE
│ ((isFree ∧
│ connect[subscriber A__subscriber__id? / caller?,
│  ring__tone / callee?] ∧
│ connect[subscriber B__subscriber__id? / caller?,
│  phone__ringing / callee?])
│    ∨
│ (isBusy ∧
│ connect[subscriber A__subscriber__id? / caller?,
│  busy__tone / callee?])))∨
│ (number__status? = INCOMPLETE ∧
│ connections' = connections)
│    ∨
│ number__status? = INVALID ∧
│ connect[subscriber A__subscriber__id? / caller?,
│  NU__tone / callee?])
└────────────────────────────────────────────────
```

```
┌─ receive__on__hook__signal ───────────────────
│ Δ (connections)
│ on__hook__signal? : Message
│ subscriber__id? : ID
├────────────────────────────────────────────────
│ ((∃connection : Connection•
│   (connection.caller = subscriber subscriber__id? ∧
│    connection.callee = callee? ∧
│    disconnect[subscriber subscriber__id? / caller?]))
│     ∨
│ (∃ connection : Connection•
│   (connection.caller = caller? ∧
│    connection.callee = subscriber subscriber__id? ∧
│    disconnect[subscriber subscriber__id? / callee?])))
└────────────────────────────────────────────────
```

APPENDIX B

Special symbols

$A \rightarrow B$ the set of all mappings which assign to each element of A precisely one element of B

$A \nrightarrow B$ the set of all mappings which assign to each element of A at most
one element of B

$a \mapsto b$ the mapping which assigns to a the element b

$\#S$ the number of elements in the set S

$\mathbb{P}S$ the set of all subsets of S

$A \cup B$ the set of all elements which belong to either A or B or both

$\cup S$ the set of all elements which belong to at least one element of S

$\text{seq}S$ the set of all sequences whose elements belong to S

$\exists$ means 'there exists'

$\forall$ means 'for all'

$\wedge$ means 'and'

$\vee$ means 'or'

$\{a,b,c\}$ the set whose elements are a, b, and c

$E[a/b,$ the expression E with all occurrences of b

$c/d]$ replaced by a and all occurrences of c replaced by d.

REFERENCES

1. Spivey J M: 'The Z notation: a reference manual', International Series in Computer Science, Prentice-Hall, 2nd edition (1992).

2. Coad P and Yourdon E: 'Object oriented analysis', 2nd Edition, Prentice-Hall (1991).

3. Rumbaugh J, Blaha M, Premerlani W, Eddy F and Lorensen W: 'Object-oriented modeling and design', Prentice-Hall (1991).

4. Shlaer S and Mellor S J: 'Object lifecycles: modeling the world in states', Prentice-Hall International (1991).

5. Booch G: 'Object oriented analysis and design with applications', The Benjamin/Cummings Publishing Company Inc (1990).

6. Martin J and Odell J J: 'Object-oriented analysis and design', Prentice-Hall (1992).

7. Jacobson I: 'Object-oriented software engineering: a use case driven approach', Addison-Wesley (1992).

8. Embley D W, Kurtz B D and Woodfield S N: 'Object-oriented systems analysis: a model driven approach', Yourdon Press Computing Series, Prentice-Hall International (1992).

9. National Computing Centre Ltd, Oxford Road, Manchester M1 7ED: 'SSADM Version 4', (September 1990).

10. Yourdon E: 'Modern structured analysis', Prentice-Hall (1989).

11. Stepney S, Barden R and Cooper D (Eds): 'Object orientation in Z', Springer-Verlag (1992).

12. Rose G A: 'Object-Z', in Stepney S, Barden R and Cooper D (Eds): 'Object orientation in Z', pp 59-78, Springer-Verlag (1992).

13. Cusack E and Rafsanjani G-H: 'ZEST', in Stepney S, Barden R and Cooper D (Eds): 'Object orientation in Z', pp 113-126, Springer-Verlag (1992).

14. Rafsanjani G-H: 'ZEST:Z extended with structuring: a user's guide', Report produced as part of DTI funded project: PROST objects (1993).

15. Cusack E: 'Inheritance in object oriented Z', in America P (Ed): 'ECOOP '91, European conference on object-oriented programming', pp 167—179, Springer-Verlag (July 1991).

16. Martin J: 'Information engineering Vol I: introduction and principles', Prentice-Hall (1989).

17. Martin J: 'Information engineering Vol II: planning and analysis', Prentice-Hall (1989).

18. Martin J: 'Information engineering Vol III: design and construction', Prentice-Hall (1989).

19. Hayes I and Jones C: 'Specifications are not (necessarily) executable', Software Engineering Journal (November 1989).

20. Wordsworth J B: 'Software development with Z', Addison-Wesley (1992).

21. West M and Eaglestone B: 'Software development: two approaches to animation of Z specifications using Prolog', Software Engineering Journal (July 1992).

22. Knott R, Krause P and Cozens J: 'Computer aided transformation of Z into Prolog', in Nicholls J E et al (Eds): 'Z user workshop', Oxford, Springer-Verlag (1990).

23. Semmens L and Allen P: 'Using Yourdon and Z: an approach to formal specification', in Nicholls J E et al (Eds): 'Z user workshop', Oxford, Springer-Verlag (December 1990).

24. Polack F, Whiston M and Hitchcock P: 'Structured analysis — a draft method for writing Z specifications', in Nicholls J E et al (Eds): 'Proceedings of the Fourth Annual Z User Meeting', York (December 1991).

25. Merad S M: 'Adding formalism to object oriented analysis', Presentation at KBSL Meeting, York (1992).

26. Giovanni R and Iachini P: 'HOOD and Z for the development of complex systems', in Bjørner D et al (Eds): 'VDM '90:VDM and Z — Formal methods in software development', pp 262-289, Springer-Verlag (1990).

27. Tse T H: 'A unifying framework for structured systems development models', Cambridge University Press (1991).

28. Rafsanjani G-H and Colwill S J: 'From Object-Z to C++ : a structural mapping', Proceedings of the Seventh Annual Z User Meeting, London (December 1992).

8

FORMAL SPECIFICATION OF MANAGED OBJECTS — A CASE STUDY

A J Judge and C D Wezeman

8.1   INTRODUCTION

BT has a programme of automation of its business processes for carrying out network, service and customer management. In pursuit of this goal, it is committed to increasing the level of interworking between the systems which currently support these processes. The use of interfaces based on open systems interconnection (OSI) systems management standards is seen as a key element in the integration of BT's systems portfolio. These standards enable management applications on different systems to interwork across a variety of communications media based on both local area and wide area network technologies. The applications are typically labelled using five broad categories:

- fault;
- configuration;
- accounting;
- performance;
- security.

If applications are to interwork they need to share a common view of the management data and functionality across a particular interface. One of the significant achievements of the OSI systems management standards work is the establishment of a common set of principles for representing the functions and data which are the common ground between interworking management applications. The standards use an object oriented approach to model this information. Coherent sets of management functions and data are represented as managed objects. The ISO standard 'Guidelines for the Definition of Managed Objects' (GDMO) [1] presents a set of information templates which can be used to define all aspects of a class of managed objects.

The GDMO templates specify both static characteristics (i.e. data typing) and dynamic characteristics (i.e. behaviour) of managed objects. The static characteristics are defined using the ASN.1 (Abstract Syntax Notation) language [2] for defining abstract data types. There is, however, no formalism, other than unstructured English text, for expressing the dynamic, behavioural characteristics of a managed object. This leads to difficulties in verifying the correctness of any implementation of a managed object specification based on GDMO, especially where the managed object has complex behaviour included in its definition.

This chapter presents the results of a study which investigated the use of a technique for formally describing the behavioural and the static characteristics of managed objects, using an object oriented variant of the Z language [3]. A further study was made of the possibility of using the formal specification directly to generate an implementation of the managed objects in the object oriented programming language C++. The aim of the work was to show that benefits can be gained from the use of formal specification techniques for defining interfaces based on managed objects. It was also intended to demonstrate that the use of recommended mappings from the formal specification directly on to a chosen implementation language can reduce the number of design decisions which need to be taken by developers when implementing managed object interfaces. Validation of these mappings prior to their use by developers should result in improvements by ensuring a close correspondence between specifications and implementations.

The work performed in the case study presented here is described in more detail elsewhere [4].

Section 2 of this chapter shows how managed objects are currently being described, and what benefits are expected from their formal specification; it also addresses existing work on specifying managed objects using object oriented Z, and on deriving implementations from object oriented Z specifications. Next, section 3 describes the method of work of the case study.

Section 4 then discusses the results of the case study. Finally, section 5 provides conclusions, and directions for future work are suggested.

8.2 BACKGROUND

8.2.1 Current practice in managed object specification

Managed objects were developed by ISO and CCITT standards groups as a means of specifying the information transferred between management systems. Management communications occur when a system:

- seeks to control another system;

- seeks to gain information about the behaviour of another system;

- needs to report an event to another system.

Each management communication involves a managing system and a managed system — either system may initiate the communication. This simple two-party communication model provides the basis for much more complex multiple-manager configurations. These are viewed as a set of two-party models.

Communications management can involve a wide range of resources, such as service level agreements, bills, racks, multiplexers or protocol machines. The characteristics of a resource may not all be of interest in management terms — an important distinction when modelling a resource is to decide which characteristics are relevant for management purposes.

To reflect its importance, the management view of a resource is uniquely identified as a managed object.

The managed object model may be used for more than simply 'physical' resources — it may represent all, or part of, a management function such as a logging or event-handling procedure. To produce managed object specifications, the resources that are to be managed must first be identified (from a requirements specification). It then remains to determine the range of management functions that each resource must make available across the interface.

A general model for management communications is shown in Fig. 8.1. The idea of a managed object as the management view of a resource is shown in Fig. 8.2, which shows the managed object as a box surrounding the actual resource with a 'window' through which management information passes.

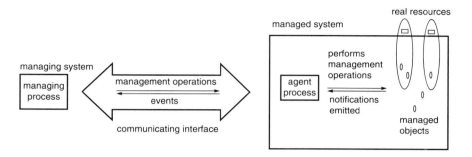

Fig. 8.1 A model for management communications.

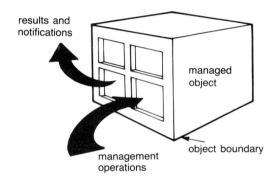

Fig. 8.2 A managed object.

An individual managed object has the following properties:

- attributes — the properties or characteristics that are visible at the managed object boundary, each attribute having an identifier and a value, which reflects or determines an aspect of the behaviour of that managed object;

- operations — the management actions which may be applied to that managed object, which are accomplished through the invocation of the appropriate managed object behaviour;

- notifications — emitted by the managed object as a result of an internal event;

- behaviour — a definition of the semantics of management operations, behaviour defines what the object does in response to particular stimuli, such as the invocation of an operation on the managed object or the occurrence of some event within the managed system which has some significance for the managed object.

A managing system can only know about a resource, or control it, through the operations and notifications that pass through the window and hence cross the managed object boundary.

Many managed objects will have the same minimum set of properties and these are said to be members of the same managed object class. The set of properties are the definition of the managed object class.

For efficiency of specification, object oriented inheritance may be used in the definition of managed object classes. A new managed object class may thus be defined in terms of an existing, parent managed object class plus a set of extensions.

For consistency of presentation of managed object class definitions, a series of specification templates are defined by ISO [1]. The ASN.1 specification language for defining data types is used to represent the static characteristics of managed objects. The current GDMO template for representing managed object behaviour is generally seen to be deficient because it allows for the use of free format English text for the definition of object behaviour. The ambiguities inherent in natural language mean that it is unsuitable as a means of expressing managed object behaviour.

At present, there are no guidelines available for implementers of managed object specifications on how to translate the object oriented concepts inherent in managed object classes into any chosen implementation language. Conventional programming languages such as C and Cobol are not well suited to the direct representation of object oriented features in code. Within BT there is increasing emphasis on the use of object oriented programming languages such as C++ to implement object oriented system designs, especially for managed object interfaces. Implementers would, therefore, benefit from some guidance on how to translate managed object designs into efficient C++ code which is also a valid implementation of the original object oriented features of the design. This should reduce the scope for introducing errors into the translation between design and implementation.

8.2.2   Expected benefits from the formal specification of managed objects

Formal description techniques (FDTs) have been sucessfully applied in a wide variety of systems and software engineering projects involving both commercial and military applications. Part of the motivation for carrying out the study reported here was the expectation that a number of benefits would result from the use of FDTs for managed object specification. The additional rigour associated with FDTs should result in a general improvement in the quality of managed object definitions. Formalization of managed object behaviour should be particularly beneficial, with consequent improvements in the internal consistency of managed object definitions.

It was also expected that conformance and interoperability testing of managed objects would benefit from improvements in the clarity of behaviour definition, with potential productivity gains from the application of automatic test case generation techniques made possible by the use of FDTs.

Finally, it is likely that there should be productivity and quality gains in the implementation of managed object definitions through the use of validated implementation mappings from the formal specification directly into an implementation language such as C++. This technique reduces the number of implementation routes from a formal specification and focuses development expertise on those aspects of a design which are not reused from earlier designs.

8.2.3   Existing work on specification of managed objects in object oriented Z

In earlier work Rudkin [5] and Aujla [6] suggested that some object oriented variant of the formal language Z would provide an excellent candidate for specifying managed objects. Rudkin [5] provides guidelines on modelling managed objects using object oriented Z, together with some examples illustrating their use. Aujla [6] shows that it is both possible and beneficial to model GDMO managed object specifications, together with their ASN.1 type definitions, into object oriented Z. He translates a GDMO managed object specification into object oriented Z, and he concludes that translation is feasible and can be simplified from that suggested in Rudkin [5].

While Rudkin [5] and Aujla [6] address how, in general, object oriented Z could be used for specifying managed objects, they do not investigate the ability of this formal language to describe complex managed object behaviour. It is this complex behaviour that is difficult to describe in natural languages, such as English. One of the aims of the study reported here was therefore to investigate whether object oriented Z can be used to describe complex managed object behaviour.

8.2.4   Existing work on implementing from formal specifications

The process of implementing from formal object oriented Z specifications has been addressed in Colwill [7] and Rafsanjani [8]; both present an (informal) structural mapping between object oriented Z and C++. They also include a series of coding conventions, which can be used to produce C++ implementations from specifications written in object oriented Z. They demonstrate the use of the mapping and coding conventions for some examples. In this study the intention is to explore whether this mapping could be used to produce managed object implementations from their formal specifications.

8.3   CASE STUDY METHOD

The study had two major aims. The first aim was to demonstrate the use of object oriented Z for specifying managed objects. The particular interest was in showing that Z could be used for expressing complex behaviour of managed objects. Secondly, if it could be successfully demonstrated that managed objects could be specified adequately in object oriented Z, the aim was to demonstrate the use of an existing object oriented Z to C++ mapping for producing implementations of formally specified managed objects.

To achieve these aims, a case study, was carried out in four steps:

- a small set of managed object specifications with complex behaviour was selected;

- an object oriented variant of the Z language was chosen;

- the chosen managed objects were specified using this variant of Z;

- the object oriented Z to C++ mapping was applied to the formally specified managed objects to produce their implementations.

The case study was performed by a team of people, which included experts in formal specification techniques and experts in managed object specifications.

8.3.1   Choice of managed objects

The case study required a choice of some examples from the numerous existing managed object class definitions. It was particularly important that the managed object definitions chosen included complex behaviour clauses. An inheritance hierarchy of managed object class definitions was selected so that the formal specification of object oriented concepts, such as inheritance, could be explored. The hierarchy is shown schematically in Fig. 8.3.

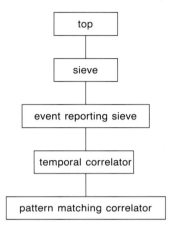

Fig. 8.3   The chosen inheritance hierarchy.

The classes 'top', 'sieve' and 'event reporting sieve' are taken from the NM FORUM Managed Object Class Library [9]. Top provides an attribute for identifying managed objects. All other managed object classes inherit from top. Sieve provides operations for reading or changing the values of attributes. Top and sieve are not used directly as managed objects in managed systems, but are defined to provide functionality for use by other, more specialized managed object classes.

The event reporting sieve inherits all attributes and operations from both top and sieve. In addition it provides a filter operation which allows it to filter out particular types of notification which it receives. This is done by matching the data in the notifications against the filtering criteria defined in the sieve construct attribute. Notifications which match positively are emitted as event reports to the address specified in the destination address attribute of the event reporting sieve.

The temporal correlator and pattern matching correlator are BT extensions to the FORUM library classes. They provide the ability to compare series of notifications (the event reporting sieve can only deal with individual notifications) using a variety of matching criteria.

The temporal correlator is a specialization from the event reporting sieve. It correlates notifications which it receives within a given period of time. The rationale for this simple approach is that many failure scenarios result in a burst of alarms being received in close temporal proximity. It also provides functions for controlling when and if notifications will be sent out as event reports. This inexpensive approach can be used in systems which are low on processing power, or as a precursor to some more complex analysis process.

The pattern matching correlator is a further specialization from the temporal correlator. It adds to the temporal correlator functionalilty, which it inherits, the ability to compare series of notifications using the values present in selected data fields in each of the notifications. The matching is controlled by specifying arbitrarily complex patterns of notification data fields which are then used as the basis for comparing pairs of notifications. For example, it can be used to correlate notifications of the same severity from the same managed object class, providing that they occur within a certain number of minutes of each other.

Another reason for choosing this particular set of managed object classes was the fact that a C++ implementation for them already existed. This was produced as part of a demonstrator system which was used for validating the specifications and as an educational tool for managed object system developers. The existing C++ implementation could be used as a basis for comparison with the C++ code generated through the use of stereotyped mappings from the formal specification.

8.3.2   Choice of object oriented Z

There exist a number of object oriented variants of the language Z [10]. For this case study ZEST was used (Z Extended with STructuring) [11, 12]. ZEST is an object oriented variant of the Z language which was developed by BT. It combines the modularity of object orientation with the precision of the formal language Z, to provide a rigorous tool for object oriented specification of complex systems. It has been developed to meet the requirements of network management systems, in which there is a continuous trend towards object-orientation and where increasing complexity requires a more rigorous approach to specification.

ZEST is very closely related to Z — and ZEST specifications can be transformed (flattened) into corresponding Z specifications. This has the benefit that existing software tools for type-checking Z specifications can be used for type-checking (flattened) ZEST specifications. To type-check ZEST specifications in the case study the type-checker 'fuzz' was used [13].

8.3.3   Formal specification of the managed objects

The guidelines of Rudkin [5] for modelling managed objects in object oriented Z, and the suggestions of Aujla [6] for their simplification, were used to produce the formal specifications of the selected managed objects. The results of the specification process are discussed in section 8.4.

8.3.4   Implementation of the managed objects

A large part of the mapping and coding conventions for transforming object oriented Z into C++ code in Colwill and Rafsanjani [7] is applicable to ZEST. This has been used in the case study for producing the C++ implementations of the formally specified managed objects. For producing the managed object implementations, use was also made of the GNU C++ compiler [14] and GNU C++ libraries [15] — these libraries were used to raise the level of abstraction of the C++ code close to that of the ZEST specifications. The results of the implementation process are discussed in the next section.

8.4   RESULTS OF CASE STUDY

The results of the case study are presented in this section. The potential advantages of the use of object oriented Z for specifying managed objects and of implementing from formal specifications are discussed. Directions for future work are also given.

8.4.1   Specification

The emphasis of the case study was to show that ZEST could adequately describe complex behaviour of managed objects. To do this the static characteristics (i.e. data typing) of these objects needed to be expressed in ZEST first, since they are referenced within the behavioural part of the managed object specifications. In the original managed object specifications, these static characteristics were described using the ASN.1 notation for defining abstract data types. In earlier work Aujla [6] had already shown that the ASN.1 data typing used in managed object specifications could be transformed into object oriented Z. Similarly it appeared that ZEST could be used to describe the ASN.1 data types of the managed objects considered in the case study.

For specifying in ZEST the dynamic behavioural characteristics of the managed objects, the guidelines of Rudkin [5] were used. Rudkin shows that a strong correspondence exists between managed object concepts and (object oriented) Z features, which can be exploited by using a series of guidelines for managed object specification in Z. These guidelines have been used as a basis for producing ZEST specifications of the managed objects addressed in this case study. In the study they have therefore been adapted for ZEST, and further guidelines have been introduced. The new set of guidelines shows how the basic elements of a managed object can be specified in ZEST. Further ZEST expressions will be needed to produce the precise description of these elements.

Figure 8.4 illustrates the use of the new set of guidelines and the further use of ZEST to describe detailed object behaviour, and also addresses an inheritance hierarchy of managed object classes which are simplifications of some of the objects addressed in the case study. It further shows how inheritance techniques used in the original description of the managed objects for creating new managed object specifications from existing ones, are mirrored by ZEST specification techniques. By using ZEST it has been possible to maintain the level of abstraction of the original, informal, specifications in the formal specifications of the managed objects.

We illustrate the use of ZEST for describing managed object classes with some examples. We define a hierarchy of three managed object classes which are simplifications of some of the classes considered in our case study. We show how managed object class concepts are expressed in ZEST, and how new classes are constructed from existing ones.

Typically a managed object class consists of a set of attributes, a series of operations and a definition of behaviour. The values of some attributes are fixed when an object is created and can not be changed. The values of other attributes are variable. Operations may be invoked by systems outside the managed system modelled by the managed object. They may involve inputs and outputs. The behaviour definition specifies what a managed object will do when an operation is invoked. It may also specify additional behaviour, which is performed by the managed object, and not invoked explicitly by another system.

New managed object classes can be created from existing classes by adding attributes and/or operations to them. Operations may also be redefined, i.e. replaced by other operations.

We illustrate the use of ZEST for describing the following hierarchy of managed object classes (the sieve and event reporting sieve classes are taken from OSI [9]).

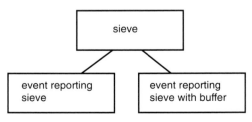

Fig. 8.4   Specifying managed objects in ZEST — an example.

First we introduce the managed object class sieve which defines a set of attributes and operations to read or change these attributes. The ZEST specification of this class is given in the box named sieve. It describes a ZEST class sieve.

The specification defines sieveId and destinationAddress to be attributes with values fixed upon object creation. Next it defines four variable attributes administrativeState, sieveConstruct, operationalState and internalState.

The variable attributes will have an initial value, which is set when the attribute is created. In ZEST, the initialization of these attributes is done in the INIT operation.

The sieve class contains a series of operations for reading or changing the values of the attributes. In the ZEST specifications below we have included two of these operations. Operation GetadminState reads the value of the administrativeState attribute, and outputs it (conventionally, outputs end in '!'). Operation SetadminState changes the value of this attribute. To do this it requires input (inputs end in '?'). The Δ symbol indicates that the value of administrativeState is changed to the value of administrativeState'.

The sieve class will contain further operations for reading and changing attributes, but they have not been included here. Also the sieve class defines some behaviour that involves the spontaneous change of one of the attributes, which we would also describe by a ZEST operation. Space does not allow inclusion of this operation either.

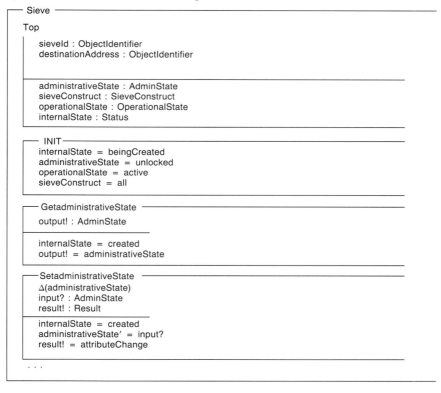

Fig. 8.4 (contd).

We use the specification of the sieve managed object class to define a new class event reporting sieve. The new class inherits all attributes and operations from the sieve class. In addition it allows instances of the class to receive alarm notifications and to forward them as event reports to some client. Notifications are received only if they pass the filter that is defined by the sieveConstruct attribute. Part of the behaviour of the event reporting sieve as described in OSI [9] is shown below.

---

eventReportingSieveBehaviour   **BEHAVIOUR**

DEFINED AS The event reporting sieve receives event data and filters it by comparing the parameters to the criteria defined in the sieve construct attribute. The event data that passes the filter criteria is forwarded to the address defined in the destination address attribute.

The destination address can only be specified when the event reporting sieve is created. Thereafter, it is a read-only attribute.

The operational state attribute identifies if the event reporting sieve is functional. Only the disabled and active values are supported. When disabled, the event reporting sieve is not functional for some reason; it is not possible for new data to be processed by the sieve. When the operational state attribute is active, the event reporting sieve is available to accept new data, process it, and forward it appropriately.

The sieving function may be suspended and resumed by management operations on the administrativeState attribute that set it to locked or unlocked. The unlocked state allows data to be sieved and the locked state prevents data from being sieved.

A value for the sieve identifier attribute can only be provided when the object is created. Furthermore, once created the value of the sieve identifier may not be modified (i.e. an instance cannot be renamed).

In ZEST we would describe the class event reporting sieve by the specification below. By including a reference to sieve all attributes and operations of the sieve class are included. The operation Filter is used to describe the behaviour of instances of the class upon receiving an alarm notification. Notice the simplicity of the ZEST specification to describe this behaviour, as opposed to the English language clause from the original specification.

```
__ EventReportingSieve _____
  Sieve
  __ Filter _____
    notification? : Notification
    output! : EventReport
    _____
    operationalState = active
    administrativeState = unlocked
    notification? passes sieveConstruct
    output! = notification?
    _____
```

Fig. 8.4   (contd.)

In our cast study we addressed much more complex behaviour, displayed by the correlator managed objects. Space does not allow this to be included here. Instead we add to the hierarchy presented here the object class event reporting sieve with buffer. We introduce it as an alternative to the event reporting sieve class, to further illustrate inheritance in object class definitions. The event reporting sieve with buffer buffers incoming alarm notifications until the buffer is full, and then outputs one event report that is derived from the contents of the buffer. To store the buffer and to record the current buffer size it uses two new attributes alarmBuffer and bufferSize. In ZEST we would describe the class event reporting sieve with buffer formally by the specification given below. It uses a ZEST operation OutputEventReport to model the spontaneous output of an event report when the buffer is full. This operation will empty the buffer at the same time.

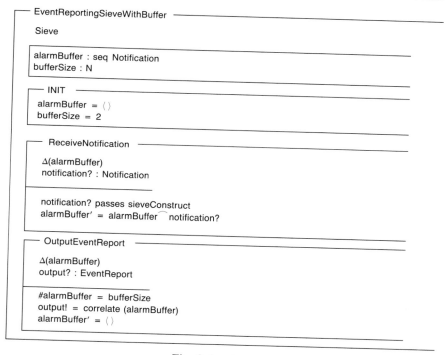

Fig. 8.4   (contd.)

The use of the 'fuzz' tool for type-checking the ZEST specifications of this case study helped in developing better and consistent specifications. To apply 'fuzz', the specifications needed to be transformed manually to Z specifications. This manual transformation was time-consuming, hence 'fuzz' was applied sparingly. For application in earnest of ZEST to specify managed objects, it is therefore recommended that a ZEST type-checker is used. No software tool existed for this purpose when the case study was performed.

To summarize, it was found that ZEST proved an extremely powerful tool for describing managed objects (with complex behaviour) formally, and hence unambiguously. ZEST brings to managed object specification the rigour of formal specification techniques. It has features that include simplicity, formality and a strong expressive power. It can be used to create specifications in a structured manner and with the level of abstraction that is required for describing managed objects. In addition ZEST supports object oriented concepts such as inheritance, that are used frequently in the specification of managed objects.

8.4.2 Implementation

Having shown that ZEST provides a powerful tool for specifying managed objects, the next step in the case study was to show that C++ implementations of the objects can be derived from their ZEST specifications.

In particular, the aim was to show that the mapping for deriving C++ from the object oriented Z of Colwill and Rafsanjani [7] could be used. The case study therefore set out to implement some of the ZEST-managed object specifications using this mapping.

The ZEST specifications provided precise and unambiguous descriptions of the static and behavioural characteristics of the managed objects. The structured approach used for describing the managed objects, proved valuable for adopting a structured stepwise approach to their implementation.

In each of the implementation steps the informal mapping and coding conventions of Colwill and Rafsanjani [7] were used. It was found that use of these conventions simplified the implementation process considerably. During the case study some of the coding conventions were improved so that more efficient code could be produced. It is expected that further improvements can be made to the mapping, and that it can be extended. This is left for future work.

The use of the mapping ensured that a close correspondence was maintained between the ZEST specifications of the managed objects and their C++ implementations. The ZEST specifications provided a very precise description of what needed to be implemented. They therefore also provide excellent documentation of the C++ code.

The implementation process was assisted by the use of GNU C++ libraries which raised the level of abstraction of the C++ implementations close to that of the ZEST specifications.

It was found that there exists a close correspondence between ZEST, C++ and ASN.1 data types. This suggests that it would perhaps be possible to create a library of corresponding types, which could be used to simplify the specification and implementation process. This library could provide a basis for creating tools that automate the process of producing data types in one language from existing data types in another language.

From the results of the case study it was concluded that C++ implementations can be derived from ZEST specifications of managed objects, and that the existing mapping will simplify the implementation process. Future work may usefully include the extension of the mapping to include more rules and existing rules may be validated and improved to produce better code.

8.5   CONCLUSIONS

This study has demonstrated that ZEST can be used to specify managed objects — especially their behaviour — in a formal manner. It has addressed successfully some of the deficiencies, such as, ambiguity and incompatibility, arising from the use of natural language to express managed object behaviour. The exercise resulted in an enhancement of the quality of the managed object classes chosen for the study, through identification of errors in their behavioural specifications. Furthermore, the use of stereotyped mappings between ZEST and C++ structures shows that it is possible to make the process of implementing from ZEST less labour-intensive than is currently the case for implementing from GDMO specifications.

These findings clearly indicate that there are benefits to be gained from using ZEST in the process of specifying managed objects. Currently, ways of making ZEST more accessible to managed object specifiers are being investigated. One method being considered is the production of standard mappings from GDMO on to ZEST structures. This would make the production of ZEST specifications from the original GDMO less specialized and time-consuming than at present. If the use of ZEST (or some related technique) became common practice for managed object interface developments, productivity gains in implementing managed object software could be realized through the use of the ZEST to C++ structure mappings. It is intended to pursue this possibility with a number of BT systems developers.

In the future, the intention is to study the possibility of automating as much as possible of the process of transforming managed objects specified in GDMO into ZEST and then transforming the ZEST into C++. The aim is to reduce the number of design and implementation decisions which must be taken by developers of new managed object classes by reusing validated paths from object oriented design into object oriented code.

Finally, it has been demonstrated that a combination of object oriented design principles with formal specification techniques can significantly improve the quality of OSI interface definitions which use managed objects.

REFERENCES

1. ISO/IEC IS 10165-4 Information Technology — Open Systems Interconnection — Structure of Management Information — Part 4: Guidelines for the definition of Managed Objects (ISO/IEC JTC1 SC21/WG4) (July 1991).

2. CCITT Recommendation X.208: 'Specification of abstract syntax notation one (ASN.1)', CCITT (1988).

3. Spivey J M: 'The Z notation: a reference manual', International Series in Computer Science, Prentice-Hall (1989).

4. Wezeman C and Anderson G: 'Complex networked systems, task 3 — deliverable 3.1, Formal Specifications of Managed Objects and their Implementations in C++', Internal BT report (March 1993).

5. Rudkin S: 'Modelling information objects in Z', in de Meer J (Ed): 'International Workshop on ODP', North-Holland (October 1992).

6. Aujla S S: 'Service creation enabler project, translating a GDMO specification into object oriented Z', Internal BT report (March 1992).

7. Colwill S and Rafsanjani G-H: 'BT corporate design project: programming from object oriented specifications', Internal BT report (March 1992).

8. Rafsanjani G-H and Colwill S J: 'From object-Z to C++: a structural mapping', in 7th Annual Z User Meeting, Wiley (1993).

9. Object Class Library Supplement: DIS GDMO Translation. Forum 006, Issue 1 (OSI/Network Management Forum) (April 1991).

10. Stepney S, Barden R and Cooper D (Eds): 'Object orientation in Z', Springer-Verlag (1992).

11. Cusack E and Rafsanjani G-H: 'ZEST', in Stepney S, Barden R and Cooper D (Eds): 'Object Orientation in Z', pp 113-126, Springer-Verlag (1992).

12. Rafsanjani G-H: 'Complex networked systems, ZEST — Z extended with structuring', Internal BT report (February 1993).

13. Spivey J M: 'The fuzz manual', (1991).

14. Tiemann M D: 'User's guide to GNU C++ (for v 1.37.1)', Free Software Foundation Inc (1990).

15. Lea D: 'User's guide to GNU C++ library', Free Software Foundation Inc (1991).

9

CONFIGURATION INTERWORKING BETWEEN SERVICE AND NETWORK LEVEL MANAGEMENT SYSTEMS

S Cairns, P H J Houseago and J R Parker

9.1 INTRODUCTION

9.1.1 Project background

There is an increasing need to automate BT's service and operational processes both to reduce operating costs and to offer a more efficient service to customers. In addition, customers increasingly seek more advanced services such as Centrex (with which PBX-like facilities are provided by a central switch, as opposed to the more conventional local switch), virtual private networks (see section 9.2), and intelligent network services. With these advanced services there is a requirement for customers to have direct control of their service features.

Typically our services are supported by a number of management systems with manual processes combining to provide the overall service. To enable improvements in service and efficiency it is necessary to automate the existing manual processes and to increase functionality at the service layer described below. In particular it is necessary to automate the interaction between the management systems.

BT's co-operative network architecture for management (CNA-M) [1], defines a structural architecture (see section 9.3) within which the processes, and hence management systems required to provide our services, are contained. The two principle layers of this architecture are the service management layer (SML) and the network management layer (NML). Interworking between these layers has been achieved using common management information services (CMIS) [2] and common management information protocols (CMIP) [3] communications and appropriate managed objects.

The service management layer includes the co-ordination of all activity associated with the management of services which BT provides, including the processes associated with the taking of customer orders, billing and fault enquiries. The network management layer includes processes by which BT plans and operates its network.

Close interaction between these two layers of the architecture is vital. Hence it is necessary to have a general-purpose interface between systems at the service and network management layers of the CNA-M. The development of such a general-purpose interface, focusing particularly on the order provision interface is the subject of this chapter. Section 9.4 discusses the customer SML interface defining service level agreements, while section 9.5 considers the SML/NML interface. Sections 9.6-9.8 discuss aspects of what is involved when a customer places an order. Sections 9.9 and 9.10 describe the NML implementation to support ordering processes and the managed objects involved. Concluding remarks are given in section 9.11.

The scenario which was chosen to demonstrate the management system interworking was a virtual private network (VPN), which is described in section 9.2.

9.1.2   Design and analysis approach

The use of an object-based interface led naturally to the decision to adopt an object oriented (OO) analysis and design method. Such an approach enabled a common analysis structure for the applications and the interface protocol.

The method is based on a number of sources — Shlaer and Mellor's work [4], the work of Rumbaugh et al [5], Booch [6], and the work on object oriented structured design notation, much of which has now been captured in the CASE tool StP (Software through Pictures), an IDE product.

The method is composed of a number of prescriptive steps within the two distinct but connected analysis and design stages. Both stages have been formalized in internal BT documents, which are used as day-to-day aids to the application of the method.

9.2   CONFIGURATION SCENARIO

9.2.1   Overview of a VPN

A VPN is a network in which customers are provided with a service that gives the impression of sole usage of a physical transmission network, which, in fact, they do not have. The owner of the network, for example BT, combines the traffic generated by many customers over the same physical transmission plant. The customer is allowed to define a coherent numbering scheme, which is composed of company-defined dialling codes, and is applied to the entire customer network. The result of this is that the network, although spread over a large physical area and carrying calls from different companies, appears to the customer as a single, very large PABX dedicated to the customer's company only (see, for example, Fig. 9.1).

The VPN modelled for this project consists of a set of customer sites called EndNodes, each with a telephony function similar to a PABX. This functionality could be provided by either a physical PABX at the EndNode, or a logical PABX provided by the network switch — i.e. a business group (BG). Each EndNode has a local dialling plan which associates a user number to the unique VPN number assigned to each EndNode. Within the VPN various EndNodes are connected by links, to reflect the customer-defined network topology and voice carrying capacity. Where no link exists between two EndNodes there is no traffic-carrying capability, and therefore no communication between them. The local dialling plan does permit unidirectional and bidirectional connectivity across the links, although for the purpose of this project, all connectivity is assumed to be bidirectional.

Figure 9.1 shows the Service_Level_Features seen by the customers at the service level, and how they map on to the features which represent them in the virtual view. Finally, these features are mapped into the logical view.

9.3   CNA ARCHITECTURE

In order that BT can have a homogeneous framework for all its communications systems it has defined the CNA [1]. This extends the principles of internal standards by specifying the agreements and specifications that enable information equipment from different vendors to be interconnected.

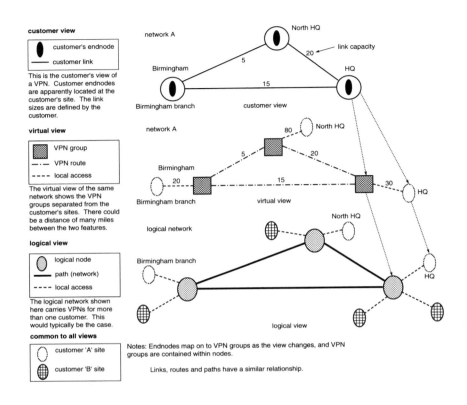

Fig. 9.1   Showing the relationship between customer, virtual and logical views.

It does this by isolating and defining the common 'veins' through the various standards — in particular the CNA-M architecture defines a management architecture with specific layers of logic and functionality.

These layers are the business management layer, service management layer (SML), network management layer (NML) and element manager layer (EML). The interfaces between these layers are ideally based on international standards, CMIS/CMIP [2, 3] and managed objects [7-9].

Further information relating to the modelling process can be found in Chapters 1 and 8.

The use of managed objects allows a common management view of the network to be used to perform the necessary control and data transfer.

9.3.1   The service management layer

The SML provides a customer oriented abstraction of network-supplied services independent of the underlying systems and technologies. In order that customers can manage their services the SML must be able to provide, create, maintain and account for those services. Analysis of these functional requirements has resulted in a process model which covers the areas of fault and configuration management for the SML.

9.3.2   The network management layer

The NML is the layer that binds the individual network elements, via their element managers (EMs), into a homogeneous network providing an information transport medium on which differing services are supplied.

The functionality offered by the network is not service related but represents a set of functional components which may be configured by the NML to provide services (see section 9.3.4).

9.3.3   The element management layer

The EML is the layer which controls individual items within the network. Each item will form a part of a service offering, but it merely offers functionality, not a service. Each element, such as a switch or a local telephone exchange, may be capable of performing more than one function, and it is the correct configuration and control of each element that provides the customer with the service required.

9.3.4   SML to NML data and object relationships

Between the various layers there is shared or 'public' management information on which CMIS services act in support of the required management actions. Isolating and defining this public information between the SML and NML defines the management interface (see Fig. 9.2).

In order that BT maximizes the usage of the capital investment in its network it must be effectively managed. To achieve this the resources of the underlying network need to be utilized in a multiplicity of services.

This is achieved by ensuring that the functionality offered by the network is not service-specific but represents a functional component to be used by the SML in providing services. The name adopted for these functional

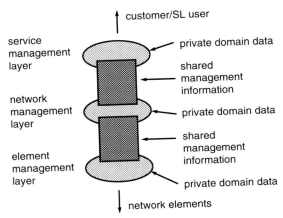

Fig. 9.2   The common shared management information between objects in differing layers.

components is Feature and the SML objects that are supported by them are SL__Features (Service__Layer__Feature).

These SL__Features represent the components of the service rather than the functionality that actually provides them. The SML has to present information in terms that are understod by a customer, therefore between the SL__Features and the supporting Features a translation is required.

This is done by the SML, as the NML is unaware of the use of its Features in providing specific service offerings. Clearly there is not a complete isolation between the layers since management actions are required and therefore this overlap of management information is maintained by the SL__Features (see Fig. 9.3), so forming the shared information between them.

The Features required are selected by the SML to meet the requirements of the SL__Feature with which the customer is interacting.

Clearly these types of inter-object relationships are not restricted to the SML to NML interface but are found across all the interfaces of the architecture as shown in Fig. 9.3.

9.3.4.1   Feature naming

The name of a Feature must be unique within a managed information base (MIB). Within the NML, there were two ways in which Features could be named:

- by the SML;
- by the NML.

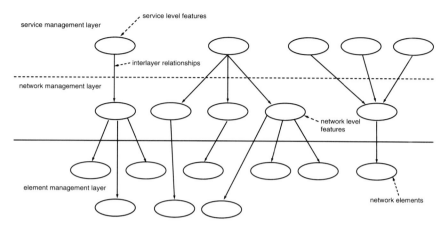

Fig. 9.3   Inter-object relationships between the CNA-M layers.

Either method will result in a name — the distinguished name — which not only uniquely identifies the object, but also implicitly gives information about the location of the object within the containment hierarchy [6] of the system.

In the former case, the distinguished name of the SL_Feature is made unique by the customer who names them. In this case, to make the Features which act in support of the SL_Features unique, the NML must also be aware of customer entities.

This requires duplication of the management information within the SML and NML, but, from its definition, only the SML is aware of customers, otherwise this would be in breach of the architecture.

Further, naming Features in this way would render impossible their use by the SML on a 'pick-and-mix' basis, i.e. the same Feature used by different SL_Features. Were a customer entity to be removed from the SML the Features associated with the SL_Features would themselves have to be removed. This would be an unacceptable restriction.

The naming of features by the NML provides a solution, in that when a Feature is created its distinguished name is determined by the NML and it is a function of the SML to maintain the relationship across the interface.

9.3.4.2   Feature inheritance

Feature objects form a 'family' of objects, i.e. siblings, subtypes, subclasses, all inheriting from the same type/class, and as such they have common attributes, some of which are:

- userLabels;

- operationalState;

- administrativeState;

- featureID.

It is apparent that these attributes may be used to form a reusable base-class from which the required specializations inherit as shown in Fig. 9.4. This follows the way that managed objects are defined [10] and that of the object oriented paradigm [5, 6].

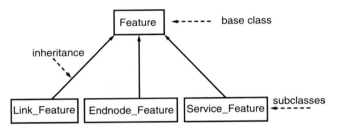

Fig. 9.4   The inheritance diagram for the feature set.

9.4   CUSTOMER TO SERVICE MANAGEMENT LAYER INTERACTION

Services are sold based on contractual agreements between the supplier and the customer with any queries or problems conducted within the framwork of these agreements. These agreements are known as service level agreements (SLA).

From a business viewpoint there is a clear requirement to reuse as many of these factors as possible with only slight modifications as required for specified agreements.

If these factors are considered as defining templates or profiles they are potentially referenced a number of times. There are several data items contained in these profiles which would not be visible or configurable by a customer. This is because the data is required purely for internal use, and hiding it from the customer prevents administrative and contractual difficulties.

9.4.1   Service level agreement profiles

Two examples of the various service profiles which are used are outlined here, but more information may be found in Cairns [11]. Details of the relationships which exist are shown in Fig. 9.5.

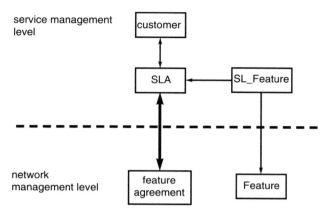

Fig. 9.5   Relationship of a service level agreement and its profiles to the feature agreement and its profiles.

9.4.1.1   Service management layer reporting profile

This profile defines the actions that are required when a SL__Feature is considered to be faulty. This may be different to the NML reporting profile since the same SL__Feature may be used by different services with different SLAs.

Some factors which would make up this profile include the time of day, the level of fault, the type of fault and the reporting period required.

9.4.1.2   Repair profile

This profile is wholly concerned with event handling and the factors identified were:

• the time by which an agreed fault should be cleared — the contract fixByTime;

- the escalation procedures which should be followed when specified time intervals have elapsed beyond the contracted fixByTime;
- the priority that a faulty Feature should assume.

9.4.2   Service management layer to network layer interaction

The SLA is concerned purely with the SML to customer interface and in order to meet these obligations the SML must be sure that the Features employed will facilitate this. The SML must have a contract with the NML via an object comparable to the SLA; this has been identified as being the feature agreement (FA). A discussion of FAs is beyond the scope of this chapter.

Each Feature used by the SML is associated with an FA with the relationship ideally being many Features to one FA and one FA supporting various SLAs. The relationship of an SLA and FA are shown in Fig. 9.5.

9.5   THE SERVICE MANAGEMENT LAYER TO NETWORK MANAGEMENT LAYER INTERFACE

The interface developed to enable communication between the SML and the NML consists of a defined sequence of messages.

Each of these messages is used to affect either the behaviour of an object, or the state of an object, i.e. the data values it holds. The affected objects are all contained within the NML, and are called managed objects. This set of managed objects allows the SML to access and alter some of the data stored within the NML.

9.5.1   Service level agreement to service relationships

There exists a defined relationship between a customer and instances of SLAs plus the services they control — some of these relationships are shown in Fig. 9.5 above. The SLA itself consists of customer-specific data, and a number of generic profiles which contain data that is service-specific. There will be profiles for service aspects such as service availability, billing and provisioning. Some, though not all, of these profiles will have network level counterparts.

The customer object must control the SLA objects in that, if the customer object is 'disabled', then all SLA instances assigned to that customer must be disabled, as well as the services that they represent. This functionality must be part of the behaviour of the objects and could require the generation of many CMIS messages to reflect the changes to the NML or to any other peer level SML.

From a customer viewpoint it may be necessary to associate many services with one SLA and to permit SLAs to interact. This permits a customer to negotiate a corporate SLA which dominates all other SLAs they may have.

9.5.2   SML to NML interface

Customer configuration control of their services may be regarded as an ordering process, i.e. the process of requesting services/goods from a supplier for receipt on a specified date.

There is therefore a relationship with the previously identified SML process of 'generic order handling (GOH)' [12].

The functional requirements of this process can be summarized as follows:

- capture the required configuration;

- permit a dialogue for the negotiation of an agreed supplier's provide-by date based on the required configuration and the customer's required-by date;

- be kept informed of the progress of the order via supplier-generated expected date of completion and order status;

- make changes to the service configuration automatically as determined by the order details upon final completion of the order;

- cancel and/or modify a previously placed order before it is completed.

In order that an SML can provide the information to meet the second and third points above, a dialogue with the NML is necessary, since only from there is the necessary access to the status of network resources available. Also accessible is the information contained within peer level network systems such as work force management [13].

Using this information, the necessary decisions as to when the required resources could be made available to meet the order requirements are made.

9.6   THE ORDERING INTERFACE

Detailed analysis of the requirements previously detailed led to the identification of three objects — the order definition object (ODO), the order control object (OCO), and the order details object (OD) within the NML.

In the present analysis, one ODO was identified. The use of an additional 'type' attribute would allow a range of ODOs, each relating to a specified set of features. Although such a possibility is recognized, it is beyond the scope of the current work.

9.6.1   The order definition object

The ODO is the unique object within the NML which exports the Feature base and the order actions (create/query/reject) to the SML. It does this through the attributes featureBase and actionsSupported, respectively.

The reasoning behind the formats and values is that it is possible for the NML to add or delete a Feature and/or action from its list as a result of some change in the underlying system.

The attributes of operationalState and administrativeState have their normal OSI/NMF meaning. In particular a busy state held within the operational state attribute identifies that no other orders can be accepted until this state changes to active or enabled.

This managed object has an operations profile which excludes the ability for a manager to delete or create the object, or set any of its attributes since the object is exporting an agent's ability to provide a specified set of Features and actions thereon; only the agent has the necessary information to indicate the respective availabilities (see Tables 9.1 and 9.2).

Table 9.1   Attributes of order definition object.

objectID	Holds the unique reference of the ODO.
featureBase	This is a multivalued attribute which is a list of the Feature supported, i.e. Group/Route VPN in this scenario.
actionsSupported	This is a multivalued attribute which is a list of the order actions supported (create/query/reject). Where this attribute details the agent's ability to support some action and goes beyond its operationalState for the object may be enabled but still cannot support a specified range of actions.
operationalState	Normal meaning.
administrativeState	Normal meaning.

Table 9.2   Operations of the order definition object.

requests	BT creates, deletes, sets, supported
notifications	attributeChange, deenrolObject, addValue, enrolObject
action types (via M-Action messages)	action types create, query, reject

The message actions (create/query/reject) support the requirements identified above to export the feature base, and the order actions. They can be summarized as:

- create — this action details that the SML wishes to place a new order with the NML (i.e. agent CME) which should then determine the effect of the order details and create an OCO with its associated OrderDetails object(s) with the NML-generated date of compliance, i.e. provideByDate;

- query — the action details that an NML is able to accept a dialogue regarding the possible modification to an existing order;

- reject — the action goes along with the query action but is used to indicate that the NML is able to not only accept a change of details against an order (OCO) but then discard the details if the SML (customer) is not satisfied with the provideByDate.

The use of the second and third actions permits a dialogue to be supported for the agreement of both details and dates.

9.6.2   The order control object

The order control object (OCO) provides an abstraction of an order placed by the SML across the interface. It does this by holding 'state' and date information (see below) about the order and forming an implicit naming path to the order details contained within the OrderDetails object(s) as shown in Fig. 9.6.

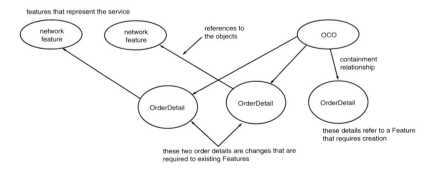

Fig. 9.6   The relationships between the OCO, OrderDetails and the Features of the service.

The OCO is generated by the NML in response to a request from the SML for a provisioning request.

9.6.2.1   The order control object states

For an order control object, there is a set of states in which it can exist. A 'state' diagram is given in Fig. 9.7.

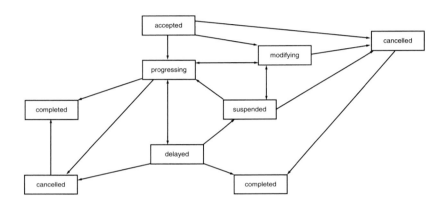

Fig. 9.7   The states of an order control object.

9.6.2.2   The interaction of the dates and OCO status

Within the OCO, three attributes hold date information — the expectedDate, provideByDate and requiredByDate.

The requiredByDate attribute holds the date by which the customer requires the order to be completed, the provideByDate attribute holds the date by which the NML has agreed to supply the order, and the expectedDate attribute holds the date by which the NML expects the order to be completed.

Unforeseen problems may cause the expectedDate to change, in which case the delayCode and delayText attributes may be used to indicate the reason for the delay.

The changing values of the expectedDate and delayCode attributes thus provide a mechanism for NML to inform the SML of an order's progress. The SML can then present the information to the customer. From a BT viewpoint, it goes a long way to providing a more flexible and responsive customer interface which the market now demands.

9.6.3   The order detail objects

There are many ways in which order details could be transferred across the interface, e.g. by the use of a complex text string consisting of 'tag and data' fields, or large text strings both held as attributes within the OCO.

Practical consideration of using such formats made it clear that it would require human intervention and would therefore be unacceptable to BT. Further, the OOA paradigm that was followed resulted in the clear identification of an OrderDetails object holding the details of the order on a per Feature basis. No 'state' attributes were identified since the object is purely holding information and not actually representing some physical 'thing' with an operational state, etc.

The OrderDetails objects consist of a family of objects, inheriting from a common base class. The base class contains the common data, such as Action — which may be create, modify or delete — UserLabels, AdministrativeState and OperationalState, with a service level reference. The derived (OrderDetail) classes will hold the attributes that are specific to the various types of order.

The number of OrderDetail classes will need to be increased as further work results in the definition of new Features. In the limit there will be a set of Features which will describe all the functionality that the NML can offer. In such circumstances it is open to question whether the use of sub-typing of the OrderDetails object is a better approach, when compared with sub-classing.

A containment relationship exists between an OCO and the OrderDetails object, as the latter will never exist without the former. Further, the semantics of an OCO dictate that it controls the order in which the OrderDetails objects hold the information.

The SML cannot directly create an OrderDetails object via the use of an M_Create service, but only via an M_Action message sent to the ODO.

There exists a very strict relationship between the various objects which are involved in the order handling process. The relationships that exist across the interface between the SML and the NML are shown in Fig. 9.8.

9.7   RELATIONSHIP BETWEEN SERVICE LAYER AND NETWORK LAYER OBJECTS

Within the SML, objects maintain the management view and the necessary control relationships over the NML order objects. The use of a single object was rejected because it centralized the control for all the SL_Features and

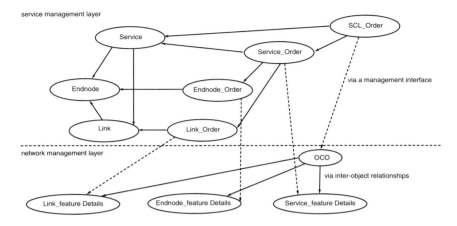

Fig. 9.8 Relationship between the objects used within the SML and NML for ordering.

the relationships to both the NML order objects and the SL__Features to be effected.

There is a clear separation between the controlling of an order's progress and the holding of the order information, since the function, and therefore knowledge required, is different.

Further, the use of a single object does not lend itself to future evolution in terms of the number of SL__Features and changes to existing ones. This is because there is no factoring of concerns towards which the OOA paradigm [3] clearly works.

These considerations led to the identification of four objects — SCL__Order, Service__Order, Link__Order and Endnode__Order.

To this end the SML object which has overall control for an order is the SCL__Order object with the same set of attributes as the OCO, and it therefore provides the same logical function.

In addition, the object also holds relationship information to the BT__Service instance to which the order applies and to the SML Service__Order object.

The SML Service__Order object in turn holds relationships to the BT__Service instance and any Link__Order or Endnode__Order objects which are related to Link and Endnode objects respectively.

All of these order-based objects hold the order data required and the action against its respective SL__Feature, e.g. Link for Link__Order (see Fig. 8).

When the 'state' of the SCL__Order becomes 'completed' the behaviour of these objects is such that they automatically perform the action required, i.e. create an SL__Feature, including database manipulations and CMIS message generation, if required.

9.8   THE MESSAGE SEQUENCES

The analysis of the information transfer across the interface and the sequence of messages to support the dialogue was constrained by the CMIS services that are available. Further details may be found in Cairns [11].

The information necessary for the order could have been transmitted via a series of M_Create messages in which the OCO is created first, followed by the OrderDetails named through the OCO.

The use of an M_Create service to transmit order information was ruled out on the following grounds:

- the supply of the order details would not be atomic and would result in inconsistencies;

- it does not permit an easy association when a modification to an existing order is required since it necessitates the creation of another OCO which 'holds' the modification details;

- it would require correlation between the existing OCO and the 'modification' OCO and this cannot easily be achieved.

Therefore another CMIS service is employed, the M_Action service, as the basis for the transport of the order details. This provides a degree of atomicity and permits a dialogue for the negotiation of dates, both against an original order and subsequent requests for modification.

9.8.1   Placing an initial order

When an SML wishes to place an order it does so by using an M_Action message sent to the NML.

The content indicates the message action (create/query/reject), the requiredByDate and the actual order details required for each Feature.

In response the NML will validate the contents and, if valid, determine the impact upon network resources to enable the calculation of a provideByDate. It will also create an OCO to represent the received order with the orderStatus attribute equal to 'accepted' and the necessary OrderDetails objects.

The NML responds with an M_Action_Result message which includes the DN of the OCO it has created to handle this order. The SML will issue an M_Get against this OCO to extract the provideByDate.

If the original message received by the NML was erroneous an error message is returned.

9.8.2   Modifying an order

A modification to an order can take place either because the provideByDate supplied by the NML in the first instance is unsatisfactory or because a presently on-going order needs to be changed in some way.

The dialogue permits the customer to issue a query against an existing order to determine the effect on the provideByDate and only when the customer is satisfied with the contents of the query, and the dates, is the OCO placed into a modify state.

In the case of the order (OCO) being originally in the state of 'accepted', the order will now start to progress. If the customer rejects the new provideByDate the SML will issue an M__Action message with an order action of reject. This informs the NML that the previous details supplied as a query against the specified OCO are no longer valid and can be discarded.

As a result of placing the OCO in the state of modify, the OrderDetails objects may be removed, created or changed and the date attributes of the OCO will also change resulting in a series of notification messages which the SML directs at the appropriate objects to maintain the required management view.

9.8.3   Multipart orders and multiple orders per service

With large configuration requests the customer would like to group various parts of the order, by date, so that they can sequence the provision. This requires a multipart order with differing requiredByDates to be sent as the original M__Action message to identify the various parts of the total order.

The relationships between the various parts of the order could clearly give rise to potential problems if a request is made for a RouteFeature to terminate on a GroupFeature which has been requested for a later date (see Fig. 9.9).

Further work is required to isolate the requirements and interactions between the various orders particularly if multipart orders are to be permitted.

9.8.4   Delayed network layer response

The previous interface descriptions support an interactive dialogue with the customer; however, in some circumstances the network layer is unable to determine the full impact of a large order in a time span which is acceptable to the customer. In this case the interface is modified in that it uses the 'enrol' notifications emitted by the OCO and OrderDetails objects.

The message sequences between the SML and NML are modified in that the M__Action__Result message in response to an order request does not

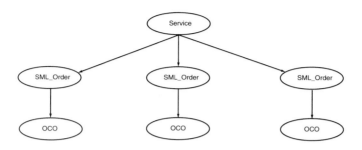

Fig. 9.9   Relationship between multiple orders and service.

include the DN of the OCO for it is to be created 'some' time later. The SML would wait for the 'enrol' notification to get the DN of the OCO and then issue an M_Get message to extract the provideByDate and slOrderRef using the latter for correlation.

Further the SML would not issue M_Get messages against the OrderDetails objects until their 'enrols' have been received, at which time it can extract the DN of the new Feature contained within the featureInstance attribute using the slReference attribute for correlation.

9.9   NML IMPLEMENTATION

9.9.1   NML facilities

The facilities required to be provided by the NML are:

- create, configure and delete network elements as required by the SML — these are requested by the SML, using a standardized dialogue that allows for an agreed provisioning date (this date is agreed after querying work-force management for the provision of new plant, if insufficient is available);

- maintain an image of the virtual network and all the associated data-fill to maintain, modify or rebuild the virtual network;

- translate its own view of the network and the service that are required into a view that the switch manager can understand;

- present and maintain a logical view of the network, as required — the NML actually uses this logical view to establish the services and features that are required at the service level;

- create, delete and modify features within the logical network as appropriate, i.e. node, path and network;

- present a service view of the network elements being used by the customer — this involves presenting the managed object view, and not necessarily a total view;

- the NML allows access for network engineering staff to reconfigure services and features as required, as well as ordering new network plant to ensure that the desired services are being provided.

In order to provide these facilities, a set of managed objects was implemented, and a communications path added. Several support processes and objects also had to be defined and implemented (see section 9.9.3).

9.9.2   NML architecture — an overview

The NML architecture, described in this section, is shown in Fig. 9.10. The major interactions between the classes are described later in section 9.10.1 and Fig. 9.14.

The NML is intended to translate the order made by the customer in service terms, into a series of instructions to communicate to the element manager, which will actually provide the service. This translation is basically from a service view to a feature view. In the context of this project, features are such network elements as routes, VPN groups, and VPNs themselves. In order to construct these virtual elements, further objects are required. These are the nodes and paths of the logical view of the network.

The view of the network is presented to the SML by means of managed objects, data from which may be passed back and forth across the interface between the two levels. The managed objects represent the only view that the SML has of the network, and the only control which exists between the SML and NML.

The purpose of the architecture, and the managed objects which can be passed over them, is to allow different views to be presented, according to the requirements of the user. One area where this is particularly useful, is in allowing the translation from the service (customer's) viewpoint, to the logical network viewpoint.

The element level network configuration is held in objects which are not managed by the SML. These maintain an image of the building blocks which

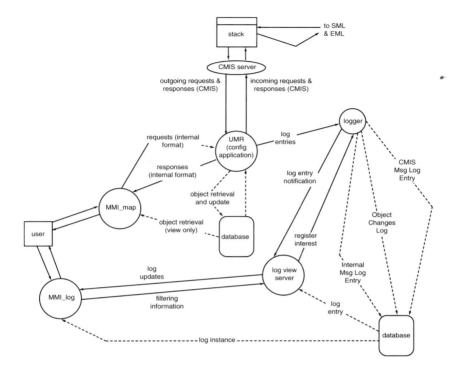

Fig. 9.10   The architecture of the NML.

the network level configures to provide a service to the customer. The NML decides how to provide the service and exactly what features to use, based upon the range of features available from the element level. These features are presented as managed objects where necessary.

9.9.3   NML processes

The major elements of the main NML processes (see Fig. 9.9) are:

- the CMIS stack — this is responsible for the communications between layers of the CNA architecture;

- the logger — this merely logs transactions for testing and audits (access paths are shown in Fig. 9.9);

- the database — this is to provide persistence (in this project an object oriented database is used);

- unsolicited message receiver — the unsolicited message receiver (UMR) is the heart of the system, being responsible for handling all incoming requests (which may be in CMIS or internal format) and all outgoing CMIS messages (which may be requests or responses) — when a message arrives the UMR itself acts on the message, retrieving objects from the database, and manipulating them or querying as required, then sending out any resulting requests and a response before it takes the next incoming message from the message queue. In order that the requests may be processed, the UMR is split into two sections:

— Order handling

This is the part of the NML that deals with the managed objects themselves, and the interface with the SML. It is responsible for the protocols used across the CMIS stack, and carries definitions of all types of orders that can be processed. It has no interface to the element manager level.

— Order processing

The main part of the functionality of the NML is to be found in the order processing sections. Here the NML communicates with the element manager levels, to configure the network in such a way that the service can be provided.

The actual configuration will be defined by the customer, working via the SML. The network level has no need to carry any knowledge of what service may be created or how; only of what is available to configure.

Similarly, the service level carries no information about the details of the construction of the service for which it has asked. The virtual components — routes, groups and VPNs — have no real meaning; at least nothing is implied about their physical realization.

9.9.4   External interfaces

9.9.4.1   Service management level interface

The service management level interface is described further in the earlier sections 9.5-9.8. The interface is conformant with the principles of CNA. The protocols used conform to the CMIP standards.

The CMIS message handler is the first part of the interface. It carries out some error checking as well as packing and unpacking of the message. The CMIS stack requires the data in a specific format, to allow further checking as it is passed across the interface.

The order definition object carries a list of all the types of order that may be executed, and will reject the order if the NML is not capable of fulfilling it. When an order is placed, this will instantiate an order control object.

This contains the necessary information to create and control an order, as well as keep track of its status. Once the order is under way, the SML will query the OCO for up-to-date status information.

9.9.4.2   Workforce management interface

There is a clear need to be able to automatically order external plant if there is insufficient available to complete an order. To achieve this, there is an interface provided, but not completed in phase I, which will be used to communicate with work-force management, to provide extra plant as required.

The interface will allow for the NML to request the plant for a certain day, such that the SML RequiredByDate can be met. As with the SML interface, there will then be provision for a dialogue if this date cannot be met.

9.9.4.3   Element management interface

The element manager interface is constructed such that it will mirror the actual network. It is not concerned with the physical realization of the features which are requested, but it does maintain a map of the logical network.

9.10   NML-MANAGED OBJECTS

The relationship between managed objects at the NL and those at the element level (EL) is shown in Fig. 9.11. Here, one SL order is controlling a single OCO in the NL, which in turn has created two OrderDetails objects. These carry the details for the two Feature objects to be created, and in their turn, create the EL configurations with managed objects at this level.

At the NML, objects are required to support the ordering interface. These mirror those defined for the SL, and are not described further. Further scenario-specific objects had to be defined to support the requirements of the project. These are the first of the set that could be defined to support an entire network.

The SML view of the network is via managed objects. These are contained within a tree, for access and management purposes as shown in Fig. 9.12.

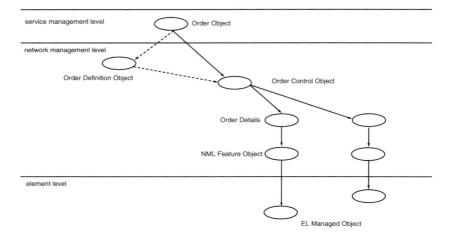

Fig. 9.11   Relationships between features and network elements.

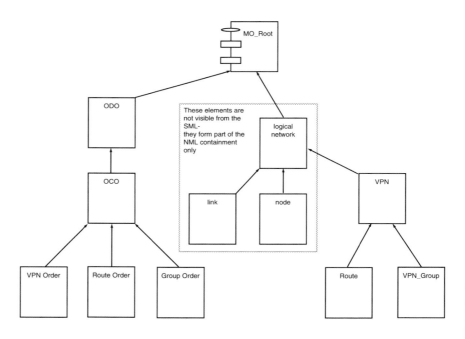

Fig. 9.12   NML containment hierarchy.

The SML has no knowledge of the logical network, as it does not use it. The NML configures components of the logical network, to provide service within the virtual network.

The first object, Route, was identified to represent the network's ability to support a voice capacity between points. This object contains attributes which relate to its service state, as well as attributes which reflect the NML view of the SML contract attributes.

The VpnGroup was identified to represent the customer's view of the endpoint. These are mirrored in the logical network by Path and Node respectively.

The networks which contain these features are identified as Vpn and LogicalNetwork; the abstractions to Connection, Terminus and Network are shown in Fig. 9.13.

Figure 9.14 depicts the interaction between the major objects in the configuration sub-system of the NML. All the objects shown, except for HandleLocalMsg and HandleCMISMsg, are database objects.

There are a number of other objects that exist within the applications on the network level which do not get stored on the database. These are

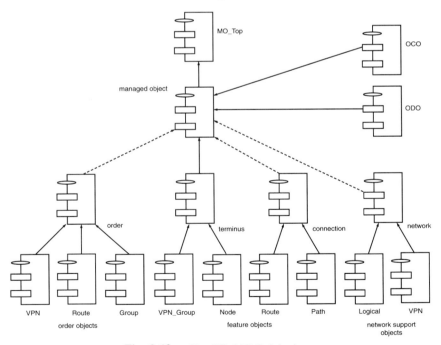

Fig. 9.13   Simplified NML inheritance.

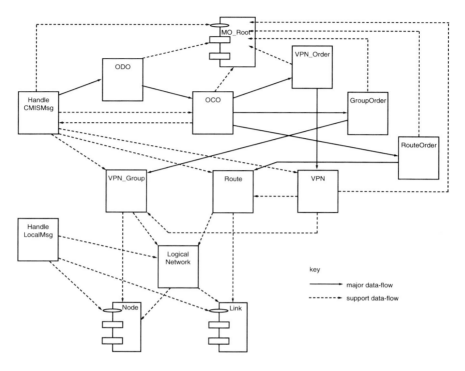

Fig. 9.14   NML usage diagram.

non-persistent objects, and they are not shown here — only the main persistent objects are shown.

Only one ODO will exist within an NML, and resides in the database. All other object classes can have many instances existing with the NML, at any given time, according to the activities within the system.

9.11   CONCLUSION

The primary aim of the work described in this chapter was to define an object oriented order-handling interface for interworking between the service and network management layers of the CNA-M architecture. This interface has been defined successfully. Furthermore it has been refined through the process of design and implementation to demonstrate its use. Work is in place to progress its adoption by international standards bodies and its use in BT's future operational management systems.

The virtual private network scenario was chosen to enable a detailed understanding of the service and network management views of VPN configuration. This demonstration of SML/NML interworking has clarified these views and led to a better understanding of the issues in managing VPN services. This understanding becomes increasingly important as BT's supply of these services grows.

REFERENCES

1. The CNA Secretariat, British Telecommunications plc, Bibb Way, Ipswich, IP1 1EQ.

2. ISO/IEC9595 Information Technology — Open Systems Interconnection — Common Management Information Service Definition (E) (1991).

3. ISO/IEC9596(E), Information Technology — Open Systems Interconnection — Common Management Information Protocol Specification (1991).

4. Shlaer S and Mellor J: 'Object lifecycles', Yourdon Press (1988).

5. Rumbaugh J et al: 'Object oriented modelling and design', Prentice-Hall International (1992).

6. Booch G: 'Object oriented design with applications', Benjamin/Cummings (1991).

7. OSI/Network Management Forum, Bernardsville, NJ USA 07924.

8. Forum Architecture 004, OSI/Network Management Forum, Issue 1 (January 1990).

9. OSI/NMF Object Specification Framework, Forum 003 OSI/Network Management Forum, Issue 1 (June 1989).

10. Forum Library of Managed Objects, Name Bindings and Attributes 006, OSI/Network Management Forum, Issue 1.1 (June 1990).

11. Cairns S: 'The service layer and its interworking', (MSc thesis) BT Laboratories, Martlesham Heath, Ipswich (June 1992).

12. Sinnadurai F N and Morrow G: 'Service management systems', BT Eng J, 10 (October 1991).

13. Garwood G J and Robinson A C: 'Work management system', BT Eng J, 10 (October 1991).

10

THE VERIFICATION, VALIDATION AND TESTING OF OBJECT ORIENTED SYSTEMS

J A Graham, A C T Drakeford and C D Turner

10.1   INTRODUCTION

The demand for larger and more complex software systems has constantly stimulated new approaches to their development. The notion of a software development process which followed a waterfall life cycle helped software engineers to understand and manage their development. Supporting this process by a structured approach to software development with methodologies such as Yourdon [1] and structured systems analysis and design method (SSADM) [2] has helped software engineers to gain greater control over larger and more complex development projects.

Recently, there has been a considerable growth in popularity of the object oriented approach to software development [3-5] and several object oriented development methods are currently in use such as HOOD [6] and Coad-Yourdon [7].

Object oriented systems are different in construction to systems developed using a traditional, functional decomposition process, and can be developed using a different life cycle [8] (see also Chapter 11). It is, however, just as important that they are subject to the same standards of quality assurance throughout their development life cycle.

Understandably, the focus of most current software testing technology is towards systems which have been developed from a functional decomposition standpoint, using a traditional waterfall life cycle and a V life cycle [9] approach to verification, validation and testing (VV&T). Verification and validation processes include reviewing, inspecting and auditing software specifications, designs and code, i.e. ensuring that developers are not only 'building the product right' but that they are 'building the right product' [10].

Testing technology will require adaptation for use within object oriented developments, and this is covered in detail in this chapter. Most of the principles and techniques described in the chapter, however, have yet to filter through to real testing environments. Verification and validation techniques, on the other hand, as they are applicable to both conventional as well as object oriented systems development, are only discussed briefly.

10.2   APPROACH

This chapter examines the suitability of the traditional V testing life cycle to the verification, validation and testing (see Wallace and Fuji [11]) of object oriented systems.

The chapter will identify where and how a VV&T life cycle can be mapped on to an object oriented development, with emphasis on the need for verification and validation throughout, and early test design. The main focus of the chapter will be on VV&T in the central part of the V model (as explained in section 10.3), where it is difficult to map object oriented development to the traditional model.

The chapter outlines the V life cycle and highlights the problems associated with applying it to object oriented systems. It then describes how to approach object oriented testing and concludes with a discussion of what should constitute a test environment for object oriented systems.

10.3   THE V TESTING LIFE CYCLE AS CURRENT BEST PRACTICE

The V testing life cycle [9] (Fig. 10.1) incorporates VV&T activities into the traditional waterfall life cycle. From the V model it can be seen that verification and validation activities occur throughout the life cycle. The tests are planned during the development of the elements of the system that they are meant to test. The V model encourages software quality assurance practices throughout the life cycle, thus ensuring that testing is not left as an afterthought at the end of implementation.

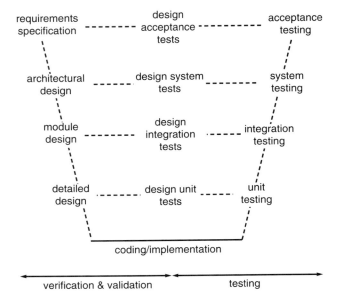

Fig. 10.1   The V testing life cycle.

The V model supports some important principles that can benefit the development by forcing the developers to consider the quality of the system being produced much earlier in the life cycle. These principles include:

- the verification of designs against specifications throughout the life cycle;

- the design of tests early in the life cycle;

- the traceability of tests and functionality throughout the designs;

- the use of designing for testability;

- the use of review techniques, walk-throughs and static analysis;

- the use of metrics to aid better design of modules and of tests.

These principles should still be recognized and used during the development of object oriented systems.

Verification and validation require that both static and dynamic techniques are used to analyse the system being developed [12]. Static techniques focus on checking the system representations such as requirement specifications, designs, and program code. This type of analysis includes:

- checking the code against coding and design standards, etc;

- checking the logic and algorithms in design specifications and in the code using code walk-throughs and inspections;

- identifying critical paths through methods (i.e. threads of control through the code) — these paths may require careful testing;

- establishing metrics on the designs or code — this is an area of ongoing research which is very relevant to the testing and test management of object oriented systems; it is, however, a relatively immature field although work is progressing on metrics for object oriented design [13].

Static analysis must be followed up by dynamic testing. This testing is performed to exercise the code at run time.

10.4   OBJECT ORIENTATION

Most approaches to systems development are based on functional decomposition, in which the system is viewed as a hierarchy of functions. However, object oriented systems are based around a different way of perceiving the development problem and structuring the system model. An introduction to object orientation is given in Chapter 1.

There are several aspects of object oriented systems which are not typically found in conventional systems. These are indicated by Booch [4] and include:

- the use of the object as the basic building block of systems;

- the instantiation of the object from a class which defines its behaviour;

- the relation of classes to one another via the inheritance mechanism.

Whilst these features are not exclusive to object oriented systems, the mechanisms used to implement these and other object oriented features, such as polymorphism, can cause problems for testers. In particular the following points may need to be addressed to fully test the quality of an object oriented system.

- The state of an object (i.e. the current combined value of all attributes in an object) is likely to affect the result of a call to its methods. However, method calls can also change the state of an object, which, in turn can affect the result of subsequent method calls. Consequently it is important to test the state of the object before and after a method call, to check that it has reached a correct state, and to ensure the correct functionality of the method [14].

- Within a class, no ordering of the invocation of methods is enforced. In theory, methods can be invoked in any order (and when the object is in any state). It is important, therefore, to test different sequences of method invocation, to ensure that methods are robust if called when the object is in an inappropriate state.

- Inheritance allows a parent's data and methods to be reused by its derived classes. This increases the number of methods, and data, that must be tested in the derived classes, and causes a large increase in the number of combinations of possible method orderings [15].

- It is important to check that the correct methods are inherited. Ideally parent classes should be tested before their children, so that some of the parent's test cases can be reused when testing the child classes [15].

- In some cases child classes have access to their parent's data. It is therefore important to ensure that there is no unnecessary interaction between the classes because of this linkage [16].

- Where polymorphism is used, testers need to ensure that the correct objects and methods are invoked.

Several other, more language-specific, issues may have to be tested for, such as the need to check memory allocation and deallocation, and the use of contracting in languages such as Eiffel [5].

If the above problems are to be properly addressed, testers must follow the principles described in the V testing life cycle, irrespective of the actual development life cycle used. The V life cycle can act as a control process that can be mapped, in part, to other life cycles in order to control the VV&T of the system being developed.

10.5   OUTLINING AN APPROACH TO TESTING OBJECT ORIENTED PROGRAMS

This section outlines an approach to testing object oriented systems. It illustrates the failure of the traditional V model and describes the need for an incremental, holistic approach to testing at algorithmic, class and object integration levels.

Testing occurs at different levels of abstraction throughout the development life cycle. The approach taken towards testing a system at a particular point in its evolution should reflect the structure of the system which is visible to the testers, at that level of abstraction.

When the object oriented nature of a system becomes apparent to the testers, the inheritance hierarchy and the relationships between interacting objects become visible. The testers are faced with two main problems:

- how to test the new facets introduced by the use of the object oriented approach for the implementation;

- how to plan the testing — the object oriented life cycle [8] involves a more iterative approach to software development than does conventional development using the waterfall model; the potential for combining object integration and class testing activities and the possible departure from a standard waterfall development life cycle make a new approach essential.

10.5.1   Acceptance and system testing

In the acceptance and system phases, the system is seen as black-box, and the underlying object structure is not visible to the testers. Consequently, traditional VV&T techniques should be applied in these phases.

10.5.2   Object integration testing

This addresses testing the system at the object integration level and considers the interaction of co-operating objects. This requires static analysis of interface specifications and dynamic, black-box testing of message passing and interaction between objects.

10.5.3   Class testing

This addresses testing the system at the class level (analogous to unit testing). It considers the interactions of methods and data encapsulated within classes, and also the testing of inheritance. It requires good static analysis of the class followed by white and black-box dynamic testing of methods, and the behaviour of methods, within objects.

Class testing also includes testing the system at the algorithmic level which considers the code within individual methods [17]. Much of the testing at this level is performed using normal white- and black-box testing (see section 10.6.1) with both static analysis and dynamic testing. It is similar to testing within units in traditional language code, with code coverage that exercises statements, branches and paths.

10.5.4 Relating class and object integration testing

Below system testing, at the class and object integration testing levels, the object oriented nature of the system becomes visible to the tester. In these levels an alternative approach to testing is needed than is currently used with conventional program testing. Figure 10.2 illustrates this central region of the V model where objects become visible to the testers.

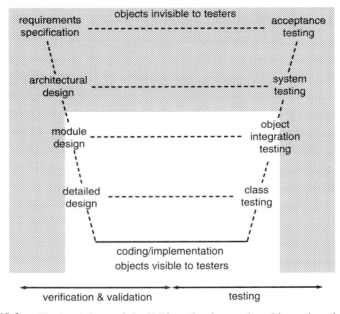

Fig. 10.2   The breakdown of the V life cycle when testing object oriented systems.

It has been argued that the class is the smallest unit of test in object oriented systems [14]. Testing at the algorithmic level is closely associated with class testing and requires the instantiation of an object and the viewing of the state of the object. Unit testing of objects can only be performed on objects that are at the lowest level (i.e. they do not instantiate other objects), or is performed on the objects that invoke stubs that emulate the objects that they would instantiate if integrated.

It is advisable to disregard the concept of unit testing for all but the lowest level objects. Instead each class should be tested in combination with, and during, the object integration testing. Object integration testing should be performed on discrete parts of the implementation as the system is built up. Class testing tests the internals of an instantiated object and inheritance, while object integration tests the instantiation and method invocation between objects.

A combination of class testing and object integration testing is needed to test an implementation and two rules of thumb can be applied to this testing [18].

- For each class that will be instantiated, parent classes should be tested before derived classes. This involves top-down testing where a parent class is tested before any derived classes. Abstract base classes must be tested by testing the first derived class that can be instantiated. Class testing is discussed further in section 10.6.

- For clusters of interacting objects, the tester should test the objects that can be instantiated before the objects that instantiate them. This is a bottom up approach and relates to object-integration testing. Object-integration testing is discussed further in section 10.7.

For example, if class *A* instantiates an object of class *B*, then class *B* should be tested first. Testing *B* will involve instantiating it and running algorithmic tests to check the logic of the methods. *B* should then be class tested to ensure that it inherits methods and data correctly from any of its parent classes (which should have been tested previously). The testing of *A* will include unit testing *A* and integration testing to check that *A* can correctly instantiate *B* and can call *B*'s methods correctly.

10.5.5   Designing for testability

Designing for testability is an important activity when developing object oriented systems, as it can ease the dynamic testing of the system later in the life cycle. The designer can help testers to view the state of objects by designing methods in pairs. For example, a method which modifies the values of an attribute can have a corresponding method which reads the attribute. This also makes the class more symmetrical, and potentially more reusable [16]. The designer can also design print methods that print out the attribute values of an object.

Another way of designing for testability is by the use of contracting (see Meyer [5]). Contracting is used to ensure that a class behaves in a way which satisfies very explicit requirements for the service which it performs. For example, a method may be contracted by a class to return a result of a certain type when it is passed parameters of a certain type that fall within certain bounds. Contracting may be implemented by the designers agreeing the contracts for each method, thus eliminating the need for code in the method to check the contract. This, however, relies on the discipline of the designers in writing code that does not break a contract.

Contracting may, however, be incorporated into the code as defensive programming. This is as a form of self-testing which is built into the system at an early stage by using invariants and pre- and post-conditions. For example, a method may check the state of the object before performing any action, thus making it more robust if called when the object is in an unsuitable state. One drawback to implementing contracting in this way is that it may cause performance problems.

The use of invariants and pre- and post-conditions can considerably cut down on later testing effort and, if the decision is taken to use contracting, it can force the designers to take much more care when designing classes.

10.6   CLASS TESTING

This section describes the key features and strategies for testing object oriented systems at the class level. It discusses class testing at the algorithmic level and strategies for testing inheritance and method-ordering effects within classes.

While full testing of a class must cover the interaction (integration) of that class's object with other objects (inter-class testing), intra-class testing focuses on validating the algorithms within each method along with ensuring that methods communicate with each other correctly. Also, the use of inheritance is checked.

It is important to ensure that polymorphism is tested by making sure that the correct methods are called at the correct time.

Within a class the methods have a slightly different nature than functions within conventional programs. Methods share access to the attributes (data) within the objects and consequently there may be a high degree of coupling between them.

As no restriction should be imposed on the order in which methods can be called, they might be called when the object is in an inappropriate state. This could cause the method to return unexpected results and a possible corruption of an object's data. It is, therefore, very important to test the order in which methods are invoked to identify problems of data corruption. This is discussed in greater detail in Turner and Robson [18].

Other language-specific measures can be taken to improve the quality of the final system. For example, in C++, a class-wide (static) variable, which is incremented and decremented when constructors and destructors are called, can be used to count the total number of instances of the class at any one time. This can be very useful in identifying memory leakage.

Memory leakage is handled in some languages with automatic garbage-collecting facilities, as with Eiffel and Smalltalk. Other languages, however, such as C++ leave memory management firmly in the hands of the designers.

10.6.1   Testing methods

Tests should be planned for individual methods (algorithmic testing) within each class. This is the lowest level of testing and involves designing both black-box and white-box tests. Black-box testing validates the external view of a class. White-box testing will validate the internal logic within methods and should include:

- identifying critical paths through methods, significant functional threads, and also critical areas that may require high priority testing;

- testing the manipulation of attribute values, by testing branches in the code;

- performing loop testing to explore the boundaries of loops;

- ensuring that destructor methods return memory to the heap.

Black-box testing should be used to test the functionality of methods within objects. Black-box testing involves validating the initial state of an object, executing a method, and validating the final state against an expected result (see Turner and Robson [18]). Black-box testing should involve:

- performing boundary value analysis and equivalence partitioning on parameters of methods within the class — this consists of identifying ranges of the parameters' values where the method should perform in a similar way, and then constructing test cases which explore the behaviour of the method on the boundary between equivalent regions, including erroneous values [19];

- testing the assertions in methods by performing boundary testing to check pre- and post-conditions;

- testing that the correct methods are called and results obtained when dealing with polymorphism;

- making sure that methods within the class can correctly call other methods in the same class, and that attributes have the correct visibility.

10.6.2   Testing inheritance

A derived class may add new methods and attributes to those declared in the parent class, or classes. The methods can be completely new, they can alter the methods from the parent class, or they can completely redefine the methods from the parent class. The methods may also access and alter data from the parent if the data is visible.

As inheritance can involve changes to attributes and methods reused (inherited) from a parent class, tests for the parent class can be selectively re-run on the derived classes object. Re-running tests in this way is known as regression testing, and is performed in order to determine whether the alterations introduced by a derived class are detrimental to the methods and attributes from the parent [15].

Tests should be analysed to see if they are still valid for the derived class. New methods in the derived class should be tested to establish that they are functionally correct and have no unexpected interaction with methods inherited from the parent class. Harold [15] has devised an incremental approach to validation of inheritance hierarchies, which reduces the number of test cases which need to be re-run.

Alternatively, the tester can visualize the derived class as a flattened class with its own data and methods, as well as the data and methods from its parents. The tester must then test the interaction of all the methods with the data, but may be faced with a 'combinatorial explosion' in the number of possible interactions.

10.6.3   Method-ordering strategies

The order in which methods are called determines the changes in state of an object. It may be possible for a sequence of method calls to put the object in a state with which a subsequent method call is not designed to work. It is useful to consider methods and data interaction [20] in terms of:

- intra-object method calling, which is the calling of methods in an object from methods within the same object;

- inter-object method invocation which is the invocating of methods in one object from methods in another.

To ensure that classes are robust, testers will need to exercise the class with sequences of method calls. However, testers will not be able to produce test cases for every possible sequence of method calls, or the permutations of values of a method's parameters. Clearly testers will need a strategy, to identify the most productive test cases, or use the approach of state-based testing (see Turner and Robson [14]), which identifies all the valid states of an object, and examines the changes of an object's state when methods are called. State-based testing, therefore, works for any method ordering.

The state of the object is defined by the combination of substates (values) of all its attributes [14]. The attributes can take on a range of values, which can be reduced to a finite number of substates by identifying ranges of values for which the methods behave in the same way, i.e. equivalence partitioning (see Myers [19]). Having identified the possible states of an object, the tester can consider the transitions between states that a method call can produce, and use this to identify where a method call will put the object in a state for which the behaviour of other methods is not defined. All methods should accept all possible states of an object.

When testing methods it is worth considering the type of each method and what it can be tested for. Methods could be tested in the following order (taken from Turner and Robson [16]).

- Constructor (or create) methods perform the initial memory allocation to the object, and initialize the state of the object. This must be tested quite rigorously by creating the object using different parameters including passing the constructor erroneous values. It is important to be able to view the state of the object immediately after initialization.

- Selector methods examine and return the current state of an object, by returning the value of the attributes examined. This should be done to view part of the state of the object after the constructor, predicate or modify methods.

- Predicate methods test for a specific instance, or type of instance, of the current state of the object by examining the value of attributes. They usually return a Boolean value. However, the state of the object may need to be viewed to check if the correct value is being returned.

- Modify methods perform a modification to the value of attributes in the object, and thus to the state of the object. Both predicate and selector methods can be used to verify the modification. Clearly, modify methods which have common attributes need to be given careful attention.

- Destructor (or destroy) methods perform the deallocation of the object — is memory returned to the heap correctly? One way to check memory allocation and deallocation is to track the amount of memory used during program execution, against some predicted norm.

The above approach is useful in making sure that methods are reused during the testing of others, and as an aid to identifying an order in which to test the methods for integration testing.

It is recommended that hidden features are tested first. Commonly a client object will invoke a method in the object under test, which will then access private data, or make a further call to a private method within the same object. Thus tests should be made on an object's private data and features by testing private methods within the object before invoking public methods by calls from other objects.

10.7   OBJECT INTEGRATION TESTING

This section discusses testing object integration by introducing the notion of an object instantiation tree which can be used to plan object integration tests. In addition, the section discusses strategies which can be used to test method-ordering effects between interacting objects.

During integration testing, the test scenarios should specify the state of the objects before and after method invocation. Consequently, in order to test a system properly, the testers must be able to view the state of the objects under test [14].

Static analysis of the interacting classes can be used to raise the developer's confidence in their quality. Typically static analysis for object integration should include:

- checking the logic and interfaces of inter-object instantiation and method invocation;

- checking that the interface is correct between instantiated objects, e.g. that parameters match, and that the calling objects have the correct private, protected, and public visibility of the attributes;

- checking that memory management is handled logically and correctly.

10.7.1   Integrating objects

Planning object integration tests can be helped by using an instantiation tree, an example of which is shown in Fig. 10.3. (It should be noted that Fig. 10.3 does not show inheritance. Also it only shows a simple tree structure, whereas the instantiation tree can be a directed graph.)

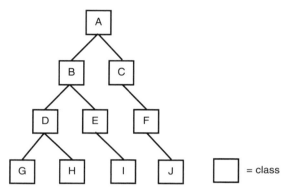

Fig. 10.3   An object instantiation tree.

The instantiation tree shows which objects can be instantiated by which other objects and how the system can grow from the top down — class *A* representing the first root object that can be instantiated when the program is first run.

The instantiation tree can be used to plan the thorough testing of the integration of the program. The testing should be performed in a bottom-up manner, where objects are tested before those which instantiate them.

Testing could proceed in the following way.

- Classes *G* and *H* are instantiated and tested first, with a combination of black-box, state-based, and white-box testing of the objects in a test harness.

- Then the branch of the tree including *D, G* and *H* is tested. *D* (which is the root of the branch) is instantiated in the test harness and tested. *G* and *H* are tested using black-box methods as objects instantiated from *D*. It may be that *G* and *H* will be instantiated several times so that several objects exist for each of them.

- Then object B is tested as an object instantiated in the test harness. It is now the root of the branch and is tested using white-box methods. D is instantiated by B and tested using black-box methods. At this point it is assumed that the branch E has already been tested in a way similar to the above, and so E is also tested in a black-box way.

- Eventually A, the root class of the program, will be reached and tested. Black-box testing of the root A can be considered as the start of the final system testing.

10.7.2   Inter-object method-ordering strategies

As discussed in section 10.6.3, the order in which methods are called should be tested both on an intra-object and inter-object basis. When testing object integration, inter-object method invocation can be used to test the interfaces and message protocols, and also highlight other method-ordering problems, e.g. ensuring that a method is not invoked in an object after its destructor has been called.

Study of the problems of testing, using orderings of methods, has identified several strategies [14, 17]. These strategies can be used to test logical threads of control through the system, or to try to ensure as much coverage of object interaction as possible:

- test pairs of features that can verify each other — features such as put and get, or write and read methods can be used so that one checks the attributes set up by the other;

- use the object life cycle strategy — this tests the object in the order 'create, modify, select, and delete', with iteration of modify and select, thus reflecting the expected life cycle of the object;

- if possible, distort the life cycle strategy — discover what happens if a method tries to select data in an object that has been deleted, or not yet created;

- critical path strategy — this takes a thread of control through the program at run time, instantiating objects and invoking methods important to the logical execution of functionality;

- tester guided strategy [20] — this is where the tester forces control along certain paths in order to test certain methods and orders of methods, and thereby the order in which objects are instantiated.

The state-based testing approach mentioned in section 10.6.3 can also be used during integration testing (see Turner and Robson [18]).

10.8   TEST ENVIRONMENTS

A test environment must be a consideration from early on in the development life cycle. In order to dynamically test object oriented programs, testers need to be able to instantiate objects and invoke the methods inside them. This means that a test environment, including a test harness, should be established so that individual objects, or groups of interacting objects, can be instantiated.

Currently, most test environments are built up in an *ad hoc* fashion as test harnesses designed to test specific code or projects. The authors are currently aware of only a few commercially available test environments for object oriented programs, other than debuggers. These include tools for testing C++ and Eiffel from MPR Teltech Ltd [21], a C++ testing tool from McCabe and Associates [22], and TestCenter, a C++ testing tool from CenterLine Software Inc [23].

A test environment can range from an individual piece of code which instantiates single objects in order to test them, to a sophisticated test harness, where many interacting objects can be instantiated working together. As object interaction is important it is necessary to test many objects instantiated together. These can then be used to send different messages to each other for testing purposes. The test environment needs some mechanism to allow messages to be sent to the instantiated objects in order to invoke the methods within them. The objects under test also need to be able to be viewed in order to see how the state of an object is being altered by the methods that manipulate attributes in the objects.

The ability to trace the flow of control through the code during the test execution and debugging may be important. The need for debugging and code analysis tools [24] means that many tool vendors are currently investigating the possibility of updating their analysis tools, especially for C++. Also some tools are being introduced for applying metrics [13] to object oriented code, but the metrics provided by these have yet to be proven.

Test environments often allow the testers to select their own paths through the code so that various orderings of methods can be invoked. These tools usually allow the tester to instantiate the object under test and emulate the rest of the system that interacts with the object. They may also allow inspection of the object under test in order to examine its state. Some tools also allow test cases to be automatically generated and stored for later use in regression testing.

In order to test objects properly the tester needs to be able to examine the internal state of an object. However, once an object is instantiated, its data-hiding and encapsulation properties limit the visibility of data. Attributes will need to be checked before and after a test is run in order to see if the values of the attributes have been altered as expected.

There are several ways to view data in an object:

- invoke those methods in an object that view, but do not manipulate, data — this is the normal way that other objects would be allowed to see data in an object, and may include the invocation of print methods that have been designed-in for testability;

- alter the private and protected domains within an object to be public (by physically altering the class representation) so that client objects in the testing environment can view the once hidden data directly;

- have direct access to the data representation — this may mean being able to view the data structure such as with a debugger, or using a trace facility (the ability to be able to view the object's state would be a desirable feature of a test environment tool).

10.9   CONCLUSIONS

The object oriented approach is gaining a growing following in commercial software development, and there has been a considerable amount of work on object oriented design and implementation. It is, however, of equal importance that effective object oriented testing techniques and tools are devised if software quality is not to be adversely affected by the adoption of object orientation.

Many of the current methods and techniques which consistute best practice for VV&T are still valid for object oriented systems. Developers still need to ensure that requirements are testable and traceable, that systems are designed for testability, and that review processes are used to verify specifications, designs and code.

Some techniques, however, such as calculating metrics on object oriented code for testing purposes, need further work before they can be applied properly to object oriented systems.

Testing occurs throughout the life cycle at different levels of abstraction. Acceptance and system testing activities should be relatively unchanged from a tester's view point. With object oriented systems special care needs to be taken at the lower levels of abstraction when the objects become 'visible' to the tester. Object integration and class testing phases need to be approached with caution as testers have less experience with this sort of testing than with conventional unit and integration testing. Also designers and implementors must give greater consideration to testability aspects during design.

Object integration and class testing needs to be differently managed to unit and integration testing in conventional systems development. Instead of occurring in consecutive steps they can occur at the same time, and a structured approach is needed to carefully test the class and inheritance testing, together with object instantiation and interaction.

A structured approach to testing the inheritance and integration aspects of object oriented programs is essential. A bottom-up approach, guided by the instantiation tree, should be used for object integration. Within classes testers should pay special attention to the state, and changes in state, of the object under test. The high degree of common coupling between methods inside classes can cause unexpected interactions between the methods, and testers must try to identify sequences of method invocation to test for this. State-based testing can be used to help to achieve this coverage [14].

In summary, everyone wants systems that are quicker to build and easier to maintain. Most of all, however, customers want systems that work, and work properly. Developers will not gain the full benefits of using object oriented approaches if they do not give careful consideration to its impact on VV&T.

REFERENCES

1.   Yourdon E: 'Modern structured analysis', Prentice Hall (1989).

2.   Downs E, Clare P and Coe I: 'Structured systems anslysis and design method — application and context', Second edition, Prentice Hall (1992).

3.   Rumbaugh J, Blaha M, Premerlani W, Eddy F and Lorenson W: 'Object oriented modelling and design', Prentice Hall (1991).

4.   Booch G: 'Object oriented design with applications', Benjamin Cummins (1991).

5.   Meyer B: 'Object oriented software construction', Prentice Hall (1988).

6.   European Space Agency: 'HOOD reference manual', Issue 3.0, European Space Agency, Noordwijik, Holland (September 1989).

7.   Coad P and Yourdon E: 'OOA — object oriented analysis', Prentice Hall (1990).

8.   Henderson-Sellers B and Edwards J M: 'The object oriented systems life cycle', Communications of the ACM, 33 , No 9, pp 142—159 (1990).

9.   NCC: 'STARTS Purchasers Handbook', Second Edition, National Computing Centre (1991).

10. Boehm B W: 'Software engineering economics', Van Nostrand Reinhold (1981).

11. Wallace D R and Fuji R U: 'Software verification and validation: an overview', IEEE Software, 6 , No 3, pp 10—17 (1989).

12. Somerville I: 'Software engineering', (4th edition) Addison-Wesley (1992).

13. Chidamber S R and Kemerer C F: 'Towards a metrics suite for object oriented design', Conference on Object Oriented Programming Systems, Languages and Applications, ACM, SIG Plan Notices, 26 , No 11, pp 197-211 (1991).

14. Turner C D and Robson D J: 'The testing of object oriented programs', Technical Report TR-13/92, Computer Science Division, School of Engineering & Computer Science, University of Durham (1992).

15. Harrold M J, MacGregor J D and Fitzpatrick K J: 'Incremental testing of object oriented class structures', 15th International Conference on Software Engineering, Melbourne, Australia (1992).

16. Turner C D and Robson D J: 'Guidance for the testing of object oriented programs', Technical Report TR-2/93, Computer Science Division, School of Engineering & Computer Science, University of Durham (1993).

17. Smith M D and Robson D J: 'Object oriented programming: the problems of validation', IEEE Conference on Software Maintenance (November 1990).

18. Turner C D and Robson D J: 'State based testing and inheritance', Technical Report TR-1/93, Computer Science Division, School of Engineering & Computer Science, University of Durham (1993).

19. Myers G J: 'The art of software testing', Wiley & Sons Inc (1979).

20. Smith M D and Robson D J: 'A framework for object oriented testing', Journal of Object Oriented Programming, 5 , No 3 (June 1992).

21. Murphy G C and Wong P: 'Towards a testing methodology for object oriented system', MPR Teltech Ltd, poster at the Conference on Object Oriented Programming Systems, Languages and Applications, ACM (1992).

22. McCabe and Associates: 'McCabe tools catalogue', 5501 Twin Knolls Road, Columbia, Maryland 12045 (1993).

23. CenterLine Software Inc, 10 Fawcett Street, Cambridge, MA, USA, 02138-1110 (1993).

24. Frakes W B: 'Experimental evaluation of a test coverage analyser for C and C++', Journal of Systems Software, 16 (1991).

Part Two

Development

11

OBJECT ORIENTED SOFTWARE ENGINEERING — PROCESS AND PRACTICE

R P Everett

11.1   INTRODUCTION

Software engineering can be considered as a number of processes which are undertaken in order to produce a final working system from some starting point. These processes are structured into a life cycle which sets out their interrelationships and interfaces. Typical life cycles include stages such as requirements capture, specification, design, implementation, testing, release and maintenance.

The introduction into an existing software development environment of a project which uses object oriented technology raises issues about process management. For example, some existing processes may not be immediately compatible with an object oriented approach. Conversely, some other processes may be used without change yet achieve equivalent targets in the object oriented environment. For a project that is going to use object oriented technology, the initiation stage is a good time at which to assess the current life cycle and the processes in it, and then decide what changes to make. This paper addresses primarily the practical management problems posed by the maturing of this new technology from a research topic into an approach which pervades all the processes of software engineering. The aim of doing this is to assist managers of future software engineering projects in fulfilling their responsibilities for the monitoring and control of all the processes involved.

The adjective 'traditional' will be used when referring to any non-object oriented process or life cycle.

In order to get some understanding of the extent of the impact of object oriented technology on the management of the processes, two projects that are using this technology were studied:

• how individual life cycle stages had been affected;

• whether the way in which the current life cycle works is still relevant to object oriented systems.

Emphasis was placed on the management of the processes rather than their technology.

The individual processes that were most changed were the central ones of implementation, design and testing, probably because they represent the most mature areas of the technology. These were also the processes which are most difficult to fit into the traditional life cycle, mainly because they span more than one stage. The following sections elaborate this theme.

The first project studied was an enhancement to a software library used for building graphical user interfaces (GUIs) which use the 'X' windowing system. The existing library, known as the InterViews toolkit, is a hierarchy of C++ classes (objects) and the enhancement consisted mainly of adding extra classes, mostly by derivation. Other changes included altering the visual styling of the components to comply with company standards. The experience gained from this long-running project forms the basis of this chapter, which concentrates on the implementation, testing, quality assurance and project management processes. The second was a project, discussed in Chapter 9, which developed a demonstration of configuration interworking between service level and network level management systems. This project was started at a time when better defined object oriented requirements capture and design methods were available. Both these projects were managed with reference to the traditional life cycle described at the start of this chapter, but with flexibility allowed in areas where object oriented technology required change to the life cycle.

11.2   REQUIREMENTS CAPTURE

At the time that these projects were undertaken, little was understood about any impact that an object oriented approach might have on the requirements capture stage. No special attempts were made to change existing techniques of interviewing, viewpoint analysis and recording the results in English text.

Consequently the management of this stage was unchanged from its traditional version. In future projects, it is possible that requirements capture and analysis processes which are specifically object oriented will become more widely understood and transferred from current research into standard practice.

11.3   SPECIFICATION

The use of an object oriented approach alters the objectives of the process. Instead of studying the requirements in order to identify functions or to identify data, the main objective is to identify objects and model their behaviour.

At the time that the requirements of the GUI product were being generated (1989), there was little information on how to do requirements analysis and there were no tools to help. There was some literature on the subject [1], and the graphical notation described by Booch was employed in the project, with the help of general-purpose graphical drawing tools. However, this product was typical in that the analysis of the requirements to generate objects was trivial, since the requirements were stated directly in the form of extensions to an existing object oriented library.

At the time that the requirements of the service level/network level interworking demonstration were being generated (1992), further literature was available [2,3] and some software tools were available. The use of these is explained in more detail in Chapter 9.

At the present time several object oriented development methods have been published [1-3] and are compared by in Chapter 7.

In view of the rapid evolution of tools and methods in this area, it is suggested that a set of 'working guides' should be established at the beginning of any project. These guides should attempt to distil pragmatic instructions based on current best practice, available tools and integration into the existing development environment. The tools so far available have been little more than customized general-purpose drawing packages. Whilst these help to illustrate object structure and interrelationships, most do not actually lead to an automated way of generating production-quality code — there is always a point at which a human has to look at the specification and the design and write the code by hand.

Recently some more complete tools have been introduced which claim a greater degree of coverage of the early stages of the life cycle.

11.4   DESIGN

The design stage is one where there is evidence of the greatest rate of change of the process. The earlier of the two projects had little access to established design processes; the latter had access both to books which explain a design process and to early tools for documenting designs; future projects can expect to have access to a wealth of information on how to do object oriented design and to better tools to support this process.

The technical aspects of design are not dealt with in this paper. However, it will be assumed that the output from it is a complete identification of all the objects in the system, including the requirements of their internal state, the externally visible interfaces, and a specification of the methods within the object. For example, in a system which is to be implemented in C++, this can be made up of a C++ class header file, plus plain text (or even mathematically formal) narrative of what the methods do.

It is at this stage that the traditional life cycle becomes difficult to work with. The reason is that after objects have been specified to this level of detail, they can be implemented and unit-tested in isolation, and that details of design, implementation and test can all be modified with little impact on other objects or the total system. This allows parallel development of objects but at the same time the development of each object spans the later stages of design, the whole of development and the early part of testing. This theme is followed up in section 11.9.

11.5   IMPLEMENTATION

The impact here is almost as great as that in design. Clearly programmers will have to be familiar with the details of the object oriented language chosen for implementation. Switching to an object oriented approach from a traditional approach almost always requires the adoption of a programming language, such as C++, designed for such an approach. This requires significantly more effort than learning the syntax of another language. Whereas changing from writing in Algol, to writing in Pascal, to writing in C, is largely a matter of learning a new syntax and then expressing old concepts in the new syntax, the move to an object oriented language requires a significant change of style of coding.

However, this shift of style ought not to be overstated. For example, the code for a method in a C++ class is reasonably similar to the code for a C function. Consequently any estimating processes (such as for productivity or fault density) which have been found to work at the level of functions can be expected to work for methods. Only when a view is taken of how the methods fit together does the difference between object oriented and traditional methods show up — but this is encroaching on the design stage described earlier.

The choice of language may in future be significant. At present, there is little choice but to use C++ for production code. This gives good efficiency and access to expertise. However, the absence of automatic garbage collection and the resulting transfer of responsibility for memory management to the designers and implementors causes a high fault rate, often discovered late in the testing stage.

11.6   TESTING

The testing process in the projects was not specifically changed to accommodate the object oriented approach. Studies have subsequently been made (see Chapter 10) of how the introduction of the object oriented approach may modify the testing process, some points of which are emphasized here.

The most important change in the dynamic testing of object oriented systems is the tendency towards a less sharp division between unit testing, sub-system testing and system testing. In their place, a process of object integration takes over. This is a spectrum of tests which starts with the testing of objects in isolation and proceeds to test progressively larger clusters of objects. Ultimately the whole set of objects is then tested as a fully-functioning system. The distinguishing feature of object oriented testing is that these tests are all conceived and managed as an evolution of the same testing activity, rather than as entirely different sub-stages.

The early stage of this evolution, object testing, is the specific instance of what is more generally called unit testing. It entails taking one object (possibly in conjunction with vestigial 'stub' objects) and subjecting it to state initialization, incoming method calls, state inspection and outgoing method call monitoring in order to provoke incorrect behaviour caused by faulty implementation.

The intermediate stages involve testing clusters of objects. For this to be feasible, there must be some cohesion to the objects so that they form a relatively self-contained functional aggregate with minimum interaction with other objects. Sometimes this clustering is a natural product of the design, but it is often possible to arrange the design to simplify the testing. This is the concept of 'designing for testability' referred to in Chapter 10. An important test that must be incorporated in all systems that use C++ for implementation is checking for memory mismanagement. The absence of a garbage collector in C++ places a heavy responsibility on the implementation and design. Typical faults include failure to release memory that has been seized, and the converse — referring to memory that has been previously released. Design approaches, coding standards and run-time checking tools are all available to minimize this problem — all are essential.

The final stage, system testing, is not fundamentally changed from its traditional counterpart. Its perception in fully object oriented systems is slightly different by being the end point of an evolutionary testing process. In many large systems, only parts are object oriented. A typical testing strategy is then to treat the object oriented part as a sub-system which is tested using object-integration testing, and then test the whole system conventionally.

Testing is sometimes taken to include 'static testing', i.e. trying to find faults without actually running code. Examples of this include code reviews, walk-throughs, and the use of tools which search code for common coding errors and poor practice. The changes to static testing involved in moving to object oriented systems are small. Extensions to checklists for reviews can quickly be formulated and updated after experience. Some commercially available tools which analyse traditional code have been updated by the manufacturers to do similar analysis on object oriented code.

The management of the testing process is therefore the planning, execution and analysis of an orderly sequence of tests on progressively enlarging clusters of objects. However, the management of the testing also has an input to the design process in order to ensure a testable design.

11.7   MAINTENANCE

This section and the following one distinguish between maintenance and support. Maintenance is the controlled change of components in response to uncovered faults or required enhancements and leads to successive releases. Support is the supply of information about the existing system (for example, by replying to queries or supplying user manuals) and provision of auxiliary material to be used in conjunction with that system; it does not primarily concern change to the system. The use of object oriented techniques tends to result in a greater proportion of customers' changing needs being satisfied by support rather than maintenance (see section 11.8). The overall effect is a reduction in cost, as the cost of producing a new release of a system to add a feature or correct a fault is vastly more than the cost of providing information about how to use the existing system to provide that feature or work round the fault.

The management of the maintenance process is not fundamentally changed by the adoption of object oriented techniques, since it still involves re-working any of the previous stages of the life cycle and planning and controlling, just as it does for traditional systems. Part of the management of maintenance is the monitoring of costs. This monitoring is important, because it becomes very difficult without it to assess whether the introduction of object oriented technology does indeed bring about the cost reductions that are claimed. The scantness of data in this area means that the claims made for cost reduction in maintenance and support must be treated as largely conjectural; an attempt is made here, however, to identify the areas where such savings might be discovered.

There is some indication that faults and enhancements are more closely encapsulated (for example, over 70% of the fixes to the GUI product required changes to only one class). That is, if a fault is reported, it is reported either explicitly against an object or against an entity seen by the end user to be closely associated with an object. Similarly, an enhancement can often be implemented by changing the source code of just one object. This again is caused by the degree of encapsulation (in the sense of local coupling between function and data) achieved by the object oriented approach. Such localization of change gives an opportunity for cost reduction by allowing reduced regression testing. The reasoning is that it may be possible to argue that localized changes cannot influence the behaviour of other parts of the system, and so some tests become unnecessary.

No close study has been made to compare the fault rates of object oriented systems with others. From what has been said above, together with the support section, it would be expected that coding faults would occur at about the same rate (expressed, for example, as faults per thousand lines of code) as for traditional systems but be easier to work round or correct.

11.8   SUPPORT

Support of a library of reusable objects poses novel problems but also provides solutions. The problems stem from the blurring of the traditional distinction between a 'library' and an 'application program'. If an object in a library does not perform precisely the task that the user of that library requires, then it is often possible to use the inheritance facility to derive a new class that provides the additional functionality. The most straightforward example of this is the addition of another method to a class, as shown in Fig. 11.1. This example, though trivial, illustrates a number of issues which occur during support. First, the turn-round time has been reduced to zero because the change has been incorporated into the user's code not the supplier's code. However, the enhancement is now owned by the user and is not available to other users, a fact which is significant if the object is intended to be reusable. Next, the question of whether this was a fault of omission in the original design is frequently a cause for debate. This is especially important if there is a contractual interface between customer and user; although the object oriented system may ultimately make it easier for the customer's needs to be satisfied, the shift of the onus of doing the work from the supplier to the customer might be a significant contractual change. Finally, if the user is expected to be able to make derivations such as this, the documentation available about the implementation of the base class must be complete and of a high standard. It may even be necessary to supply the source code. Points such as these and many more need to be carefully addressed when devising support contracts and service level agreements.

```
class A {int i, j; int Sum() {return i + j;}}
```

```
class MyA:A{int SumSq(){return i*i + j*j;}}
```

Fig. 11.1   Adding an extra method by derivation — the supplied class (top box) library has a class called A with a method Sum. The user wanted to add a method called SumSq. Rather than changing the library, the normal procedure is to derive a new class and define the new method in the new class (bottom box).

11.9   LIFE CYCLE MANAGEMENT

The life cycle used above is one which is in common use during the development of network management software within BT. However, it is worth considering whether this could be improved.

Experience shows that strict adherence to linear application of stages of design, implementation and testing does not make best use of object oriented technology. Section 11.4 showed that at the end of design and the beginning of implementation there is a clear set of strongly decoupled objects which may be implemented, unit tested and corrected independently. A rigid interpretation of the traditional life cycle would prevent rapid development of the fully functioning objects since it suggests that implementation would have to be complete before unit testing could be done. Furthermore, changes to details of the design of objects after unit testing would require extensive change control overheads, since they would be deemed to be changes to the design. What is needed instead is a structure which allows rapid iteration around the stages of final design, implementation, and unit testing. Finally, the natural parallelism in the development of the objects could not be exploited — the life cycle suggests synchronized development of all the objects.

What is advocated in its place is a life cycle in which the later stages of design, the whole of implementation, and the first part of testing (unit testing) are perceived as a number of parallel development processes, referred to here as 'object development'. This switch from sequential stages of the traditional life cycle to an iterative parallel development is illustrated in Fig. 11.2.

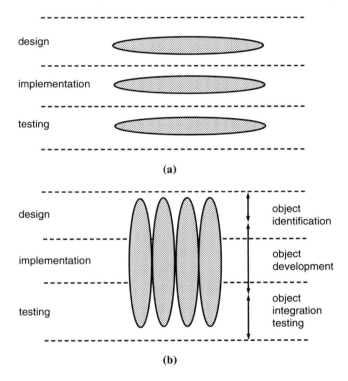

Fig. 11.2 A typical traditional life cycle (a) separates out design, implementation and testing. In object oriented development, objects undergo an iteration involving the later stage of design, the whole of implementation and the first part of testing. Stretching implementation to cover these activities confines the iteration to one stage called here 'object development' (b), where objects develop independently, leading to parallel development.

Monitoring the progress of development can then be based on the aggregate of the status of each of the objects, where the status is an estimate of how much more iteration around the design/implement/test loop is needed for that object.

The parallelism during object development, which started during design, is ended during object integration testing, which is the gradual process of aggregating the objects to make the entire system.

11.10   BUSINESS EVOLUTION

The adoption of object oriented approaches to software engineering is claimed to be one way of achieving software reuse. However, this reusability is not achieved unless the reusable component is designed for reuse and the subsequent projects actively pursue reuse targets. A particular problem is the conflict during initial projects between the need to get the product completed in a short time with low costs and the need to generate reusable code. The benefits of reuse are reaped only during subsequent projects. Consequently a business which wishes to exploit reusable code has to take a long-term view of the costs and benefits. Figure 11.3 shows qualitatively the expected results. The dotted line shows the approach to completion of a non-reused design. B1 shows how the first project which attempts to generate reusable classes may take comparably longer; but subsequent projects

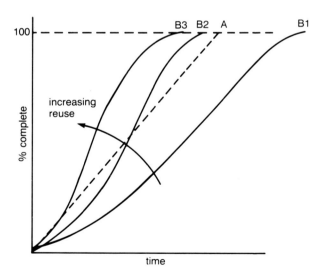

Fig. 11.3   Showing the hurdle of reuse. Suppose the system implemented in a non-reusable manner approaches completion as shown by the dotted diagonal line leading to A. Then, if the system had been implemented with reusable components, this would have taken more time (B1). However, subsequent similar systems (B2, B3 ...) can exploit the reusable components and so be implemented in a shorter time.

(B2, B3...) then save time by reusing the components, so that the business gradually benefits from the build-up o reusable components.

When object oriented technology is applied to the issue of reusability, the result is families of reusable object libraries, the composition of which reflects the real world objects of the business. Careful management is required to ensure that such a library is constructed correctly and is built up as an asset and not treated as a maintenance liability. The introduction of object oriented techniques alone will not automatically achieve such a goal.

Even if such business-wide reuse (reuse 'in the large') is not a target for a project, then within one development project there is often scope for reuse of code (reuse 'in the small'). At the project management level this is substantially the same activity as factoring out common subroutines was in traditional development. Provided the reusable entity, whether it be object or function, is identified before implementation, incorporating its development and use by the other components is straightforward. However, it is not unusual for common components to emerge during implementation. The mechanism for dealing with this situation can be coupled into the life cycle advocated here, because the explicit iteration during implementation allows for objects to be re-implemented using the newly discovered common factor.

11.11   CONCLUSIONS

This chapter has been based on a retrospective look at two projects which have been undertaken using object oriented technology in order to assess the changes needed to the management of processes that make up present-day software engineering. The process of requirements capture is not yet fundamentally changed, although research in this area may percolate into common practice. The subsequent processes leading through analysis and into design are changing, and whereas recently there was little information to draw on, future projects will have access to better defined processes of object oriented development. It is in this area that design tools are making some impact, and the present state of the art provides specialized graphical editors together with extra design checking features. The coding process is changed but to a lesser degree. Unit testing is simplified because of the tighter encapsulation achieved by an object oriented approach. All testing stages are likely to become merged into an evolutionary stage known as 'object integration testing' which will be harder to manage than strict division into unit, sub-system and system testing. Maintenance and support is technically easier but harder to manage. Finally, the life cycle presented at the beginning can be improved in order to describe more accurately the desired processes.

REFERENCES

1. Booch G: 'object oriented design with applications', The Benjamin/Cummings Publishing Company Inc (1990).

2. Schlaer S and Mellor S J: 'Object life cycles: modelling the world in states', Prentice-Hall International (1991).

3. Rumbaugh J, Blaha M, Premerlani W, Eddy F and Lorensen W: 'object oriented modelling and design', Prentice-Hall International (1991).

12

A MANAGED OBJECT MODEL OF GENERIC SWITCH SERVICES

P V Muschamp

12.1   INTRODUCTION

This chapter describes the work performed to define a managed object model for the configuration of services on digital switching elements (for related managed object model work, see Chapters 8 and 13). The definition of this model has been the result of structured analysis and specification of the interface between support systems and network elements in the domain of configuration management. Specifically, the analysis has been concerned with provision, modification and cessation of customers' lines and the services which may be provided on those lines.

The chapter is organized as follows:

- section 12.2 describes the background to the analysis work with reference to the current operational environment in which BT performs customer configuration;

- section 12.3 describes the approach taken in the analysis through the use of a simple example;

- section 12.4 describes the way in which the analysis was taken as a basis for the definition of the managed object model;

- section 12.5 outlines the ongoing work in this field by describing a practical application for the model.

12.2 BACKGROUND

12.2.1 Current operational environment

With over 25.5 million customers, BT has one of the largest networks of any of the world's telecommunications providers. With such a large customer base, the management of the network elements needs to be undertaken in a controlled environment in which support staff perform their tasks with the aid of automated support systems. The requirement for automated support systems includes the need for a centralized operations and maintenance function to manage the local and trunk switched network.

Since 1985, BT has introduced the Operations and Maintenance Centre (OMC) to its Network Operations Units throughout the UK. These systems presently serve the inland customers on all-digital switching elements. The functions of the OMC include facilities to add and remove customers, change the services provided on customers' lines, read customers' records from an on-line database, and automatically allocate equipment resources. A simplified system schematic is shown in Fig. 12.1 and illustrates the functions concerned with the configuration of customers and their line services. Support staff enter requests for service creation, modification, or cessation at user terminals. Requests are processed by the system which may call upon automatic assignment components for equipment allocation.

Each OMC will communicate with a number of digital elements, normally contained within the same geographical area. These switching elements (often referred to as switches or exchanges) are supplied to BT by a number of different vendors (see Chapter 13 for procurement information). The vendors differentiate their products from those of their competitors by offering switches capable of supporting a larger number of exchange lines or different functionality as well as on quality of service, cost, etc. The hardware and software architecture for a switch supplied by one vendor may differ radically from that for a switch supplied by a different vendor. This includes the way in which support systems such as the OMC communicate with the switches through the communications language, commonly referred to as the MML (man/machine language). As a result, the OMC needs to understand the MML for all the switch types with which it communicates; the OMC

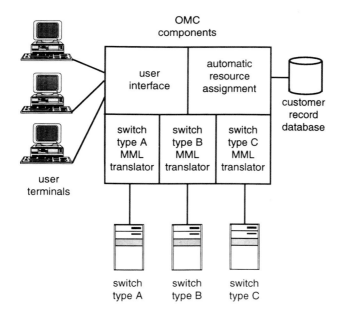

Fig. 12.1   OMC system schematic.

components, shown in Fig. 12.1, include a number of MML translators. Each request for service will use one of the MML translators to automatically construct the commands necessary to communicate with the appropriate switch.

The use of this system within BT's network has enabled the successful transition of customers on to newly-installed digital switches. Over 70 OMCs located throughout the UK are responsible for the transmission and processing of a total of more than 12 million messages (transactions) per year.

12.2.2   Introduction of new services

As described in section 12.2.1, the OMC software architecture is centred upon a number of components which perform service request and allocation functions and which pass control to the MML translators that construct the switch commands. For each switch type there exists a distinct set of translators which have knowledge of the proprietary MML commands. OMC users access a common set of data entry screens, regardless of the switch type on which the user is requesting service.

As the switch vendors upgrade their products, enhancements to the OMC software, including the MML translators, become necessary. Minor changes to the translator software can be accommodated easily due to the organization of the OMC's software components. In the case of major changes, however, the requirements for upgrades to the software become more demanding. Moreover, the introduction of a new switch type into the network with a different MML requires a new translator component to be developed and installed in the OMC. The decision as to whether to fund such development will be largely based upon the estimated revenue-earning capability of introducing the new functionality into the network. However, in some cases, even where the development is warranted, the time-scales involved in such a large development activity may prove to be prohibitive.

A number of switch vendors have the capability to supply proprietary support systems which are geared specifically to the management of the switch type supplied by that vendor. Typically, these systems provide lower functionality than that provided by the OMC and are of use in the management of one switch type. However, where the business wishes to install a new switch type which is not supported by the OMC, the purchase of proprietary support systems with the switches themselves may provide a cost-effective solution. This has a number of implications:

- the support system developments would not be under the direct control of BT, with the result that changes to the systems (enhancements, problem correction, etc) would need to be carefully managed with the vendors;

- the intellectual property would not be owned by BT with the result that the systems could be sold to competitors, thus reducing the business's ability to differentiate its services;

- the support systems may be disparate from the set of systems in use in the BT network with many implications for network operations.

In cases such as this a choice must be made between costly and time-consuming developments to the present support systems and the purchase of disparate, proprietary systems which may cause operational problems for the business.

The introduction of a standard, generic interface for the communication of commands between support systems and network elements (such as digital switches) would enable changes to network element functionality to be accommodated more easily. This is possible because the resultant changes necessary to the support systems themselves can be accomplished in a more controlled environment. More importantly, the introduction of a new switch type could be accomplished with minimal change since the services provided by the new element can be configured using a set of communications messages common to all.

The operational use of such an interface requires BT and the switch vendors to develop applications on either side of the interface which are capable of communicating using the generic message set. In BT's case, this means replacing the OMC translator components with a single component which translates requests for service into common messages which can be understood by all the switch types. The inclusion of new switch types into BT's network in the future may be dependent on them supporting a standard, generic interface, since these types can be easily accommodated into the network with minimal change to the support systems. This is essential if BT is to be able to respond rapidly to its customers' needs.

12.3   CONFIGURATION INTERFACE ANALYSIS

12.3.1   Analysis scope

Section 12.2 described the need for a standard, generic interface for communications between support systems and network elements. The scope of data which may be sent across this interface encompasses a number of functional areas including configuration management, event management, test management, performance, billing and call record data.

The area of configuration management includes customer lines and services, equipment provision, routeing table management, and fields traditionally associated with the data which is initially built into the switch database when the system is commissioned. Whilst the management of all this data is key to the correct functioning of the network, the field of customers' lines and services is one which is characterized by a high degree of activity between the management systems and network equipment. The development of efficient applications to support open communications in this field can result in considerable operational cost savings for a large telecommunications provider such as BT.

The configuration of services on digital switches covers customers whose lines terminate on a variety of equipment types. Most of the customers in the UK whose lines terminate on digital switches fall into one of the following categories — analogue, ISDN, and Centrex (the provision of PBX-like services on main network switching elements). It is this set which forms the basis for the analysis described in the remainder of this section.

12.3.2   Service example — call pick-up

The specification of a message set which defines a standard, generic interface requires analysis to be performed of the capabilities of all the digital switch types in the BT network. These capabilities include establishing and removing customers' lines and applying one or more features, or services, to those lines. One means of assessing a switch's capabilities is to analyse the MML commands used to communicate with the switch. The commands give detail of the low-level data items which, when collected together in a data transaction and sent to a switch, define the precise functionality of each service.

One example of the services which can be provided against lines connected to a digital switch is call pick-up. The service is typically used in an environment in which a number of telephones are situated in an enclosed area and allows users to answer calls ringing at other telephones. Two types of call pick-up are available:

- directed call pick-up — users enter a short code on their telephone followed by the directory number at which the call is ringing;

- group pick-up — users enter a short code to pick up a call ringing at any directory number in a pick-up group.

In order to provide this service on a customer's line, a set of commands must be supplied to the switch on which the customer's line is connected. As decribed in section 12.2.1, the precise form of the MML commands which are used will depend upon the switch type. The commands contain lists of data parameters which the switch needs in order to provide service. Figure 12.2 shows the sets of commands needed to provide the call pick-up service on two different switch types.

```
switch type A: SUBCH <DN> <EN> <TL>
               SUBSUS
               SUBSS <CPU> <CPT=2>
               CPGRP <3014>
               SUBEN
switch type B: ALTER
               RESOURCE <DN>
               SERVICE   <PICKUP>
```

Fig. 12.2   Call pick-up MML commands.

In each case, the MML consists of a number of command keywords (e.g. SUBSUS, ALTER) followed by appropriate data parameters (e.g. DN, CPT = 2).

For switch type A, five commands are used:

- SUBCH — change the customer identified by the data DN (directory number), EN (equipment number), and TL (tariff level);

- SUBSUS — suspend the data for the customer;

- SUBSS — provide the service given by the data CPU (call pick-up), and CPT = 2 (the type to be provided is group pick-up);

- CPGRP — the customer is a member of the call pick-up group identified by 3014;

- SUBEN — enable the data for the customer.

The data supplied to the switch consists of parameters which uniquely identify the customer followed by data concerned with the service. However, the following commands and data parameters are redundant or duplicated information:

- EN — the switch database contains the data which associates directory numbers with equipment numbers and can provide this value from that database;

- TL — the tariff level is constant for a given customer and would be provided to the switch when the customer is initially provided with a line;

- SUBSUS, SUBEN — the suspension and enabling of the customer's data are actions concerned with the architecture of the switch software and are required for all data changes;

- CPT = 2 — the command CPGRP identifies that the customer is to be a member of pick-up group 3014. This implies that the service to be provided is group pick-up rather than directed pick-up.

The minimum set of data which defines the request for service for this customer is thus:

- DN — which uniquely identifies the customer;

- CPU — which tells the switch which service to provide;

- CPGRP 3014 — which defines the call pick-up group in which the customer is to be included.

For switch type *B*, three commands are necessary:

- ALTER — change the data for a customer;
- RESOURCE — change the data associated with the customer identified by the directory number DN;
- SERVICE — provide the service PICKUP (call pick-up).

There is less redundancy associated with these commands since only the ALTER command can be dispensed with. However it can be seen that switch type *B* does not provide the means to allow the customer to be a member of a call pick-up group, i.e. only directed call pick-up is available as a service with this type. In order to construct a set of generic service commands for call pick-up the following two rules must be applied:

- the commands must be capable of specifying the full range of service features available on all switches, i.e. a generic service will provide the super-set of functionality offered on the switches — in this case the functionality includes both directed and group pick-up, despite the fact that switch type *B* does not support group pick-up;
- the commands must consist of the minimum set of data items needed to specify the required functionality, i.e. the redundant items must not be included — in this way, the generic message set gains in efficiency over each of the proprietary message sets.

Applying these rules to the above example for call pick-up, the essential data is the following:

- a command to tell the switch that the call pick-up service is required;
- the directory number to which the service is to be applied;
- the type of pick-up required (group, directed, or both);
- the group number for membership if appropriate.

12.3.3   Generic switch services

The example provided above illustrates the way in which the MML commands are analysed for a simple service in order to specify the set of data items which can be used to define a standard, generic message set for that service.

The approach described is applicable to all the services provided on all the digital switch types used in the BT network including those which are available on analogue, ISDN and the Centrex lines. The result is a list of commands with their associated data parameters which can be used to define a standard, generic message set, able to be used to configure all switch types.

In addition to detailing the services available on customers' lines, analysis is required in order to construct a valid set of generic data parameters to establish the lines themselves and to associate directory numbers with the correct equipment in the switch. In the case of a single analogue or ISDN line, the process is similar to that carried out for the line services, i.e. the set of MML commands and data parameters required for each switch type is compared, the redundant or default parameters discarded, and a generic set defined.

In the case of a Centrex line the process of line provisioning is more complex. By definition, Centrex is concerned with the provision of PBX-like services on main digital switches.

Part of this service is the association between the lines in a Centrex group for call routing. In addition, there must be the means to associate attendants (operators) with Centrex lines within the group as well as certain services which are applied to the group as a whole.

The approach taken for Centrex introduces the concept of a hierarchy of data. At the upper-most level the data associated with the provision of the Centrex group itself is defined (dialling plans, common services, etc). Below this level the data associated with each Centrex line within the group is defined (directory number and equipment numbers, line terminating equipment type, etc). Each line inherits the data associated with the Centrex group.

By using this approach, the commands and data parameters required to configure a Centrex group for each switch type can be analysed in the same way as the analogue and ISDN line services described above.

The generic requirements description document [1] describes the analysis approach further and defines the complete set of generic services available on analogue, ISDN and Centrex lines. The approach is essentially 'bottom-up', in that the low-level detail of the switch functionality is abstracted into a consistent set of services with the redundancy stripped away. This ensures that the set of generic services covers the functionality presently offered by digital switches.

The following section describes how the generic service information is transformed into a formally-defined model which can be used as the basis for open communications.

12.4   MANAGED OBJECT MODEL SPECIFICATION

Each of the generic service descriptions given in Muschamp and Rutter [1] is defined in terms of the behaviour of the service (i.e. the functionality it provides) and the data parameters needed by a switch in order to provide the service. The document uses English as the specification language. In order to define a generic, standard interface capable of use on all switch types, a more formal means of specification must be used.

Object oriented software engineering methods provide a well-structured means of turning English language service descriptions into a formal set of message primitives. These methods have been used to specify the configuration of customers' lines and their services in the form of a managed object model, the benefits of which are described in Chapter 1. This is illustrated below by considering the call pick-up service example described in section 12.3.2 and how this is turned into managed object specification.

12.4.1   Managed object example — call pick-up

A managed object can be considered to be a data structure which defines the behaviour of a real-world entity, from the perspective of the management of the data within it (for further tutorial information, see Chapter 1). Real-world entities include the services which are offered to customers on digital switches. In order to turn the informal definitions of services into a formal managed object model, a modelling technique must be employed. The International Standards Organisation (ISO) defines a set of guidelines for the definition of managed objects [2]. These guidelines formally define the way in which the information is to be specified and describe the way in which managed object classes can be related to each other where the management of a given object has impact upon the management of a related object.

An example of a managed object class definition for the call pick-up service is given in Fig. 12.3. This shows the use of the formal syntax specified in the guidelines [2] to define the functionality (given in the behaviour definition) and the data parameters which the switch needs to provide the service (given in the attributes). Additional information is needed to define the attributes and their permissible values and to define the relationships between this object class and others. For example, it can be seen that the definition does not contain a reference to the directory number of the customer. This reference is described elsewhere in the managed object model as a relationship between this managed object class and a directory number managed object class.

```
callPickup     MANAGED OBJECT CLASS
DERIVED FROM   customizedService;
CHARACTERIZED BY  callPickupPackage  PACKAGE
    BEHAVIOUR       callPickupBehaviour    BEHAVIOUR
    DEFINED AS
    !Allows the user to answer a call ringing at another DN.
    Varieties include directed pick-up, where the user enters
    a code followed by the DN from which the call is to be
    picked up, or group pick-up, where the user enters a code
    to answer any call ringing on a DN in the pick-up group.!
    ;
ATTRIBUTES
        cpuType              GET-REPLACE,
        cpuIdentifier        GET-REPLACE
    ;
REGISTERED AS   <package id>;
;
REGISTERED AS   <object id>;
```

Fig. 12.3   Call pick-up managed object class definition.

Using these techniques, a managed object model (a set of class definitions and their relationships) has been defined for the whole range of generic service data associated with the configuration of customers — the switch subscriber configuration managed object model (SSCMOM) [3]. Due to its formal nature, this model can be used as a basis for open communications between support systems and digital switches using the appropriate protocols.

The model defines a standard interface specification which enables BT to rationalize the data configuration process, as outlined in section 12.2.2. Vendors who supply switching systems capable of supporting the standard interface will still be able to retain a competitive edge, but this will need to be on the basis of functionality, i.e. even though the interface is standardized, the scope of functionality is not standardized and is able to be exploited by the vendors. The scope of the managed object model may need to be expanded over time to cope with new functionality provided on the switches. However, the use of an object oriented software engineering approach enables these changes to be accommodated more easily.

12.4.2   The customer administration model

The European Telecommunications Standards Institute (ETSI) has established a working group to define standards for the management of customers on digital switches. This includes customers taking service on analogue and ISDN lines. With participants from all the major European telecommunications providers and element vendors, the working group has defined a managed object model which is generic for the provision of service in this area. This model is known as the customer administration model (CAM).

The CAM is the result of a number of years of effort on the part of the participants and provides an efficient structure for use as the basis for communications between operational support systems and digital switches. The analysis approach used within ETSI follows a 'top-down' methodology, in that the model is based upon a set of standard service offerings. However, the SSCMOM has a wider scope than the CAM, in that it includes Centrex functionality. In order to take advantage of the super-set of functionality offered by both models and the complementary nature of the analysis approaches, the extra functionality provided in the SSCMOM has been merged with the CAM and agreed within ETSI.

The resultant model contains managed object class definitions of the type given in section 12.4.1 for each entity required for the configuration of service on digital switches. This includes objects which define the switch resources (such as directory numbers) as well as objects which define services (such as call pick-up). Central to the understanding of the relationship between resources and services is the entity relationship diagram (ERD). This provides a diagrammatic illustration of the interaction between managed object classes in the CAM. The switching service fragment from the ERD is shown in Fig. 12.4.

Each managed switching element (switch) contains one or more directory numbers (DNs) and customer profiles. Each customer profile represents the information for a given customer connected to the switching element and is associated with one or more DNs and access ports. These define the switching element resources which provide the customer with the means to make and receive calls.

The services, such as call pick-up, are defined by sub-classes of the customized service object. They are contained in each customer profile and are related to the resources which provide them through the customized resources object.

It can be seen that an instance of customer profile may contain one or more other instances of itself. This is the way in which Centrex groups and lines are included in the model. The Centrex group data will be defined by a sub-class of customer profile and will contain a number of Centrex lines, each of them being a further sub-class of customer profile. This is a formal representation of the hierarchy of data referred to in section 12.3.3.

The present issue of the customer administration model represents the culmination of the analysis and specification of a managed object model for generic switch services. The model is presently an Intermediate European Technical Standard [4].

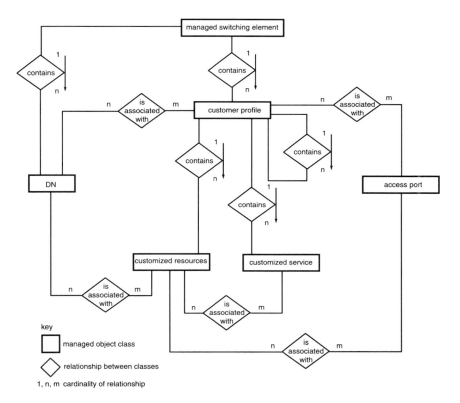

Fig. 12.4 ETSI CAM ERD, switching service fragment.

12.5 PRACTICAL APPLICATIONS

The analysis and development of a standard, generic interface specification gives telecommunications providers the means to use open communications between support systems and digital switches. This is heavily dependent upon the switch vendors developing applications on the switches which can enable communications of this type. The vendors who are the first to develop such systems can look to an increased market share from all the major service providers.

The telecommunications providers themselves need to fund developments to support system applications which enable open communications. Whilst the benefits of open communications are clear, the funding of such developments needs to be supported by a strong business case which can provide operational statistics which show cost savings for the business.

BT has developed a prototype system which tests the practical issues concerning open communications between support systems and digital switches. The scope of this development includes the two main functional areas associated with manager/agent communications of this type, namely configuration management and event management. The development is being undertaken in collaboration with a switch vendor who has supplied a prototype application to support the communication of messages to and from the system using the message primitives defined in the ETSI customer administration model.

Further collaboration in this area will address issues associated with the operational use of an open interface. Included in these is the issue of performance. With an annual message rate of over 12 million transactions, the OMC presently provides BT with the ability to manage its inland customers in a cost-effective manner. By using a more efficient message set, such as provided by the ETSI CAM, and by taking advantage of technological advances, it is expected that the future support systems will be capable of much greater transaction rates with consequent operational savings for the business.

12.6   CONCLUSIONS

The analysis and specification of a managed object model of generic switch services has been performed as a result of the desire to create a standard, generic interface for the configuration of customers on digital switches. By integrating the model with a European standard, the scope of the model has been enhanced to cover all the needs of the participating telecommunications providers and element vendors.

It is expected that telecommunications providers will increasingly seek to procure switches whose functionality includes the ability to support a standard interface. This will enable communications between the operational support systems and the switches to be performed on a non-proprietary basis. Correspondingly, support systems will be developed which will contain only a single set of communications primitives — using the interface standard as a basis for specification.

The ETSI customer administration model is believed to offer an advanced and stable specification for the management of customers and their services on digital switches. As an Intermediate European Telecommunications Standard, it is expected to be used by telecommunications providers as the basis for communication with digital switches and may become the means by which the functionality of procured switching systems is specified.

Trials based upon the use of the customer administration model for switch configuration have been undertaken and show the management of customers' lines and their services using open interfaces. Ongoing work with the trial systems may form the basis for the development of operational support systems. BT will then have achieved a level of vendor independence such that elements can be procured on the basis of quality of service, cost, functionality, etc.

REFERENCES

1. Muschamp P V and Rutter P: 'Generic requirements description document', Generic NLC Analysis and BT development Project, Internal BT document (March 1992).

2. CCITT Recommendation X.722|ISO/IEC 10165-4: 'Guidelines for the definition of managed objects', (June 1991).

3. Muschamp P V: 'Switch subscriber configuration managed object model', Generic NLC Analysis and Development Project, Internal BT document (March 1992).

4. Intermediate European Technical Standard DE/NA 943309: 'Functional specification of customer interface on the OS/NE interface', (1993).

13

NETWORK ROUTEING MANAGEMENT SYSTEM — A PRACTICAL EXAMPLE OF AN OBJECT ORIENTED DEVELOPMENT

C J Selley

13.1   INTRODUCTION

The network routeing management system (NRMS) is a software development for the BT department which operates and maintains the public switched telephone network (PSTN). At the outset, the project customer sets a number of objectives against which the success of the development would be measured. Since the NRMS is intended to replace a number of existing network management support systems, the project objectives centre on delivering:

- significantly reduced life cycle costs;

- significantly improved ease-of-use.

The specific objectives of the project are discussed in more detail in section 13.3.

The project team identified that object oriented development techniques offered the potential to make significant progress in achieving these objectives. In collaboration with the internal BT customer, the decision was taken to apply object oriented (OO) techniques to the analysis, design and implementation of NRMS. The publications by Coad and Yourdon [1, 2] provided the basis for the team's understanding of the OO paradigm and of the stages involved in OO analysis and design. All implementation work was carried out in 'C++' and the application was given persistence through the use of the ONTOS object database management system [3, 4].

This chapter describes the background to the project objectives, the development approach based on OO principles (including an example of OO design within NRMS) and presents evidence to date on the achievement of the project objectives. NRMS provides an interesting case study into the value of object orientation in operational software developments, since it is a replacement for existing systems. Hard evidence is now available on the relative costs of similar enhancements to systems developed using traditional Yourdon techniques, and to NRMS developed using OO techniques.

13.2   OVERVIEW OF THE TELEPHONE NETWORK

The BT public telephone network (PSTN) comprises the following two principal elements (see Fig. 13.1).

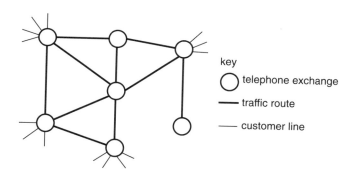

key

○ telephone exchange

— traffic route

— customer line

Fig. 13.1   Network schematic.

- Telephone exchanges

  Telephone exchanges may host customer telephone lines, i.e. they may support a direct connection between the customer's telephone and the PSTN. They also support a number of traffic routes between exchanges, such that calls from a customer hosted on one exchange may be passed on to other exchanges in the network, ultimately to arrive at the exchange hosting the called customer.

In addition, exchanges within the PSTN may support traffic routes to/from the exchanges of other network operators (e.g. Mercury).

- Traffic routes between exchanges

  The exchanges represent the nodes within the network. Nodes are interconnected with traffic routes, which carry telephony 'traffic' generated by customers hosted on a particular node, to other exchanges within the network.

The internals of a telephone exchange comprise a large number of 'routeing resources' which handle the call, manipulating the called digit string, and redirecting the call onward through the network via one or more traffic routes to distant exchanges, or directing the call to a hosted customer's line.

The source data which describes the current configuration of routeing resources is large in volume (approximately 1 Mbyte of ASCII data per exchange), complex to interpret and exchange-type specific. The source data from a System X exchange can only be interpreted by a reader with detailed knowledge of System X terminology and the function of each type of routeing resource. Similarly, for another exchange type, 'AXE10', the company maintains a skilled workforce with detailed knowledge of the 'internals' of AXE10 exchanges.

Routeing changes are made to telephone exchanges on a regular basis. The business drivers which lead to changes are fourfold.

- Network modernization

  BT has a rapid programme of network modernization, in which groups of analogue exchanges are replaced by a new digital exchange. The introduction of new exchanges impacts upon the routeing configurations of other exchanges in the network.

- New network strategies

  BT is continuously reviewing the topology of the network (i.e. what is connected to what) to offer increased service resilience for customers. Topology changes impact upon the routeing configurations of exchanges in the network.

- Introduction of new services

  BT is continuously introducing new services with specific telephone codes, e.g. a new international code for a country now accessible by direct dialling. New services are accessible from all BT exchanges and hence their introduction impacts upon the routeing configurations of all exchanges in the network.

- Support for interconnection with other network operators

  Other network operators introduce services based on digit codes. Such services are accessible from the BT network and hence impact upon the routeing configurations of exchanges in the network.

In summary, routeing configuration changes take place on a daily basis on BT's network of 3000 exchanges. Each exchange is programmed independently of all others (see Chapter 12).

13.3   NRMS OBJECTIVES

13.3.1   User requirements

The fundamental user requirements for NRMS centre on the provision of a suite of tools for:

- viewing the current configuration of routeing resources within the exchanges of the PSTN;

- auditing existing PSTN routeing configurations against routeing rules — such rules represent a standard set of routeing configurations for a large proportion of all calls, guaranteeing a known standard of resilience to network failures;

- planning the introduction of new routeing configurations in support of new or existing services.

Each of these facilities is dependent on the provision of a routeing model of the PSTN, built from information taken from the individual exchange databases. The model of the network is the core around which applications for the planning and auditing of routes may be built.

13.3.2   Shortcomings of existing systems

In the past, support systems have been produced to assist in viewing the routeing resources and the configuration of these resources within System X exchanges. They suffer from three characteristic shortcomings highlighted by the customer:

- they address only System X exchanges, and yet the PSTN comprises numerous exchange types;

- they present the user with a view of exchanges which is reliant on a very detailed knowledge of System X — its proprietary data structures and terminology (such systems are only usable by certain personnel with a highly detailed knowledge of System X);

- they are expensive to maintain.

Each of these shortcomings aligns closely with the three objectives of the customer in investing in the development of the network routeing management system.

- The PSTN comprises two principal digital exchange types — System X from GPT and AXE10 from ETL. In the future, the PSTN may include other exchange types.

  The PSTN connects with other BT networks, which carry traffic for particular services. The advanced services unit (ASU) network uses Northern Telecom exchanges, and the digital derived services network (DDSN) is built from AT&T exchanges. In the international network, BT has Northern Telecom, AT&T and ETL exchanges. A major constraint on BT in selecting new exchange types for the PSTN is the non-availability of network management support systems which are exchange-type independent. A major objective of NRMS lies in demonstrating that it can support multiple exchange types, and furthermore that it can be enhanced at minimum cost and in minimum time to support any additional exchange types which may be adopted in the PSTN.

- With the existing range of mainly exchange-specific network management support systems, the cost of operating the PSTN is high; BT must maintain specific workforces for the management of the System X network, the AXE10 network, the TXE4 (old analogue exchanges) network, etc. The customer requires that the network routeing management system should present a layered view of the PSTN to users. At the outer layers, the exchanges in the network should be presented as generic exchanges, with generic exchange features and capabilities, and at lower levels they should be presented with exchange-type specific detail. The concept of layers of abstraction would facilitate the management of certain aspects of the PSTN, without exchange-type specific knowledge. Clearly, this would offer considerable flexibility to BT in managing the network with a single workforce.

- Existing systems which model the routeing capabilities of the PSTN (System X subset) are subject to regular enhancement. At approximately six-montly intervals, GPT releases a new version of the System X software for deployment to the network. New software introduces new routeing capabilities and these capabilities must be modelled in the existing support systems. The customer's objective for NRMS was to reduce the cost and time taken to make the necessary changes to the PSTN routeing model. The achievement of this objective was imperative since NRMS would be deployed initially with two exchange types — System X and AXE10 — implying twice the number of enhancements each year and potentially twice the maintenance cost of existing System X-only support systems.

13.4   THE APPROACH

The key characteristics, which defined the OO approach (see Chapter 1), are fivefold.

- **Abstraction**

    Abstraction is the principle of ignoring those aspects of a subject that are not relevant to the current purpose in order to concentrate more fully on those that are. This principle has been applied with rigour to the analysis and design of the network model for NRMS. The exchanges which form the network are highly complex hardware/software systems. The proprietary data structures which describe their configuration for a particular instance of exchange are similarly complex; furthermore the data structures of a System X exchange appear to bear no relationship to those which describe an AXE10 exchange.

    The task of the NRMS team was to build a layered abstraction of exchanges, which at a high level would describe generically what an exchange does with respect to routeing. The layers of abstraction would reveal, at a lower level, how a specific type of exchange (System X or AXE10) achieved each generic function.

    At a high level of abstraction, an exchange may be described as follows:

    'An exchange is a node in the PSTN which receives telephone calls from local customers or from other exchanges and forwards these calls either to other local customers or onward in the PSTN to another exchange'.

This generic description of an exchange applies equally to System X and to AXE10 and for that matter to any other existing or future exchange type. An object class Exchange exists within the NRMS network model. It has few attributes — a name, a signalling address, a type — and it has a few abstracted services, such as RouteCall. It is a generic class, from which other specialized classes are derived. It is highly stable — no matter what new and sophisticated routeing services future releases of System X software may bring, this class is unlikely to change.

At a lower level of abstraction, the 'what' must be expanded to explain the 'how' for a given exchange-type. At some stage in the design process, generic abstractions specialize into exchange-type specific object classes which mirror more closely the proprietary design detail of a particular brand of exchange. These low-level classes are more volatile and more susceptible to change as the detailed functionality of the exchange is modified and enhanced from one release to the next.

- **Encapsulation**

  Encapsulation (or information hiding) is the principle of hiding a single design decision behind a defined interface. In OO the unit of encapsulation is a class — a cohesive data set collected together in a single class with a defined set of operations/services which are the only public means of accessing the data. The NRMS model includes approximately 80 classes, initially designed and documented using the Coad and Yourdon diagramming techniques and subsequently implemented in C++. The selection of appropriate encapsulations is vital to an elegant analysis and design. Class selection is achieved through use of layered abstraction in describing the required system and in utilizing inheritance structures wherever possible.

  Successful encapsulation achieves localization of the parts of a sytem which are most susceptible to change as requirements are changed (minimizing cost of change) and also simplicity, by virtue of the fact that it minimizes communication between the parts of the final system (see Chapters 1 and 6).

- **Inheritance**

  Inheritance is a mechanism for creating new classes by taking an existing class and adding to it new attributes or services. It allows the analyst or designer swiftly to create powerful and complex new classes from existing classes. It captures commonality between classes and caters for a single and reusable design/implementation of that commonality.

Inheritance structures form the backbone of the NRMS network model design. The design is principally a series of layered abstractions expressing the workings of distinct parts of an exchange. Classes at the top of the NRMS inheritance hierarchies are generic to all exchange types, and those at the bottom are typically exchange-type specific.

- **Message-based communication**

  An object oriented system is populated with instances of defined classes. Class instances communicate via message-passing. A message is an imperative form of communication — an order from one object instance to another object instance to perform a given service. Message-passing, in the general case, is a two-way communication — a requestor initiates a request and a responder provides a reply.

  Messaging-passing is the mechanism by which a static object design is animated — in the NRMS network model, object instances do nothing until they are requested to perform a service via a request in the form of a message. An object oriented system responds to external stimuli through co-operation between class instances, based on message-passing. In testing the NRMS design, prior to implementation, the technique of examining the flow of messages between object instances, in response to an external stimulus, was employed. This is a major area for CASE tool support in the future.

- **Assembly structures**

  A common feature of the NRMS network model design is an organization of class instances into assembly structures, mirroring collections of like and unlike instances under an umbrella object. Assembly structures allow the whole-part relationships, by which a telephone network is described, to be expressed as a set of exchanges and routes between exchanges. All three are abstract classes, and the former is an assembly of sets of the latter two.

  Collection classes will appear in all designs and, for NRMS, extensive use was made of C++ collection class library members such as sets, lists and dictionaries. Any aggregate structure within the PSTN domain was modelled using 'assembly structures' within this design and implemented using existing collection classes available from the ONTOS class libraries [3, 4].

These key characteristics of OO are widely recognized and well documented by many authors. In themselves they are useful but they do not constitute an analysis and design method. They do not provide an adequate framework ⸱within which to progress a project development. For this framework, the NRMS team used the guidelines provided by Coad and Yourdon [1, 2]. Coad/Yourdon does provide a set of ordered activities by which an analysis may be progressed and guidelines on the means of testing the result.

In practice, the method is loose and the documentation and testing of analysis and design results is hampered by the lack of suitable tool support. The NRMS network model analysis and design is described through the production of Coad/Yourdon diagrams with a standard drawing package.

This approach is in contrast to work on other projects where the rigorous use of StP (Software through Pictures) [5], Teamwork or other CASE tools is a well-established part of analysis and design practice.

The use of Object International's CASE tool, OOATool (PC-small project version), was discontinued at an early stage when it was found to be cumbersome to use and deficient in its automated checking facilities. In particular, the OOATool does not permit a designer to follow a series of messages between classes, to test the dynamic behaviour of the design. A second tool, P-Tech, was evaluated briefly and despite offering the promise of attractive functionality, such as model population and animation, was not adopted, since its diagrammatic conventions did not map easily on to the Coad/Yourdon conventions familiar to the team members.

The process of analysis concentrated on modelling 'what' exchanges do. The temptation is to dive into the low-level detail of particular exchanges and accurately model those of a System X or AXE10 exchange which have finally been understood after lengthy consultation sessions with experts in each of the exchange types. Building a layered abstraction of an exchange was a time-consuming and costly process.

In order to build abstractions which are truly generic, it is vital to understand, at a minimum, two examples of proprietary exchanges. Hence, all members of the NRMS analysis/design team now understand both System X and AXE10 exchanges. That approach is inherently inefficient — four people have learnt about two exchanges in considerable detail. Following a traditional analysis and design approach, the team would have been split — one group addressing System X and the other, AXE10. The analysis and design would have been completed faster and more time would have been available for implementation. The product of the in-depth, two-exchange study was the body of knowledge necessary to build a layered abstraction of PSTN exchanges, identifying commonality where it exists and isolating exchange-type specifics where necessary.

Of all the classes defined for NRMS, 70% are generic. This indicates the scale of reuse achieved within the system. The cost was incurred in a lengthy analysis-and-design phase and the pay-back is made during implementation and during subsequent maintenance. Fewer classes are faster to code. High-level abstract classes are highly stable and low-level abstractions are most volatile.

Within the NRMS design, exchange features which are most susceptible to change are encapsulated in a known set of classes. Assessing the impact of a known change is relatively straightforward and the scale of the change in the model is likely to match the scale of the change in the real-world exchange.

13.5 PRACTICAL EXAMPLE

In this section, a practical example of design within the NRMS network model is explored. The selected example is the design of the 'digit decoder', a component of a generic exchange identified during analysis and subsequently designed in detail. This section will present an analysis of digit decoder for System X and AXE10. It will demonstrate a traditional approach to design based on entity relationship modelling and the OO approach to the same design.

The digit decoder component is described as follows:

'The digit decoder is that component of an exchange which operates on the incoming digit string (often the dialled digit string from the customer), manipulating the digit string where necessary (inserting/removing digits), identifying calls to hosted customer numbers and differentiating calls into groups according to the received digit string, before passing the calls to other parts of the exchange which deal with routeings to local customers and routeings to other exchanges. The digit decoder also identifies invalid calls of all kinds; those that are to non-existent numbers (for which a 'number unobtainable' tone is appropriate) and those for which a recorded announcement is appropriate.'

13.5.1 System X digit decoder

The data which describes the configuration of a System X exchange is captured by running a GPT decompiler tool against an archived copy of a particular exchange database. The decompiler produces tabular ASCII data which is 'human-readable'. The following five points provide a description of the table types and the associated decode algorithm which represent the System X digit decoder.

- Any incoming digit string (from a local customer or from another exchange) is matched against each of the tables BA400, BA403 and BA416. The key into each table is a compound of the digit string (e.g. '0717287561') and a point of entry value. Different 'points of entry' are configured to deal with calls originating from different parts of the network. For instance, different points of entry will be configured for local customers sited in different BT charge groups and yet more points of entry will be configured for calls entering the exchange from other exchanges in the network.

- The incoming digit string is matched against the longest matching sub-string found in any one of the three tables (e.g. 071728). There is no means of predicting in which of the tables a particular digit string will be found.

- If the digit string matches on the BA400 table, then the digit decoder has completed its work — the received digit string is verified against 'minimum' (MIN) and 'maximum' (MAX) permissible length values found in the table entry and the call is passed to a 'routeing category' — a routeing resource (identified by routeing category number) in another component of the exchange (see Fig. 13.2).

point of entry	digits 1 2 3 4 5 6 7 8 9 10 11	total digits		routeing category number
		min	max	
163	7 1 7 2 8	10	10	312
163	7 1 7 2 9	10	10	312
163	7 1 2 6 1	10	10	367
163	7 1 2 6 2	10	10	390
163	8 1 1 2	10	10	25
163	8 1 1 3	10	10	26
164	8 1 2	10	10	48

Fig. 13.2   Example BA400 data.

- If the digit string matches on the BA403 table, a 'code translation number' is found and this provides an index into a code translation table, BA421. The index will match a 'code translation digits' entry (see Figs. 13.3 and 13.4).

  Simply, the received digits are now discarded and replaced by the new 'code translation digits'. The work of the digit decoder re-starts with the old point of entry and the new set of digits.

- If the digit string matches on the BA416 table, a 'retranslation number' (RetransNumber) is found and this provides an index into a retranslation table, BA423. The index will match a 'digits' and 'point of entry' entry.

| point of entry | digits | code translation |
	1 2 3 4 5 6 7 8 9 10 11	number
163	1 0 0	14
163	1 4 2	15
163	1 5 1	25
163	1 9 2	15
163	1 9 3	46
163	1 0 0	14
164	1 4 2	15

Fig 13.3   Example BA403 data.

code translation number	code translation digits 1 2 3 4 5 6 7 8 9 10 11
14	9 3 4 2 5
15	9 3 6 2 7
25	9 1 2 3
46	9 1 2 4

Fig. 13.4   Example BA421 data.

Of the sub-string of digits which matched to a particular entry in the BA416 table, a number are retained, given by KeepDigits and the string found in the PrefixDigits field is prefixed. The newly-formed digit string is re-applied to the digit decoder at the new point of entry, indicated in the BA423 table entry (see Figs. 13.5 and 13.6).

point of entry	digits 1 2 3 4 5 6 7 8 9 10 11	retranslation number
163	4 7 3 6 3	22
163	4 7 3 6 4	22
163	5 0 1 2	25
163	5 0 1 3	25
163	2 3 4	46
163	2 3 5	46
164	2 3 6 2	46

Fig. 13.5   Example BA416 data.

retranslation number	point of entry	prefix digits 1 2 3 4 5 6 7 8 9 10 11	keep digits
22	27	1 4 7 3	1
25	848	5 0 1 2	2
46	237	2 3 3 4	3

Fig. 13.6   Example BA423 data.

13.5.2   AXE10 Digit Decoder

The digit decoder functionality of an AXE10 exchange is captured in a 'print' from the relevant exchange. The AXE10 offers a set of commands (a human-computer language) for configuring the resources of the exchange. The exchange responds to commands by producing human-readable prints.

In order to examine the manner in which incoming digits are manipulated within an AXE10, an engineer will issue an 'ANBSP' command and the 'B-number analysis data' print will be produced by the exchange (Fig. 13.7).

B Origin—digits		num mod	B Origin	route	charge	length	account
253 –	0						
253 –	06						
253 –	060						
253 –	0602						
253 –	06023		F = 329		CC = 1	L = 10	A = 1
253 –	060237	M = 4 – 23					
253 –	062						
253 –	0623						
253 –	06237						
253 –	062372						
253 –	0623720		F = 332		CC = 231	L = 10	A = 231
		M = 4					
253 –	0623721		F = 332		CC = 231	L = 10	A = 231
		M = 4					
253 –	0623722		F = 332		CC = 231	L = 10	A = 231
		M = 4					
253 –	0623723		F = 332		CC = 231	L = 10	A = 231
		M = 4					
253 –	062375		F = 332		CC = 231	L = 10	A = 231
		M = 4					
257 –	1121						
257 –	11217			TE			
257 –	1125			RC = 19		L = 16	
257 –	1127		N = 259			L = 25	
257 –	1128			RC = 28		L = 7	
265 –	2						
265 –	25			ES = 102			

Fig. 13.7   Example AXE10 'B Origin' print.

Any incoming digit string (from a hosted customer or from another exchange) is matched against the second column of numbers in the ANBSP print. The first column of the print contains B Origin values. 'B Origins' are used to distinguish between calls from hosted customers and calls from other parts of the network. A typical exchange configuration might show a dozen or more distinct B Origins. Hence the key to the ANBSP print is complex — B Origin plus incoming digit string (see Fig. 13.7 above).

The call processing implied by the ANBSP print is as follows:

- locate the segment of the ANBSP print with the appropriate B Origin for the call under analysis;

- match the incoming digit string against the shortest sub-string in the print;

- read any parameters in subsequent columns and carry out any call processing implied;

- match the incoming digit string against the next-shortest sub-string in the print;

- read any parameters in subsequent columns and carry out the processing implied;

- repeat the previous two steps until a termination condition is reached.

At the fourth and sixth steps the parameters are interpreted as follows:

- $F = 322$ — a termination condition implying 're-apply the full incoming digit string to the digit decoder at a new B Origin' (in this example, B Origin 332);

- $N = 259$ — a termination condition implying 're-apply the remaining (undecoded) incoming digits to the digit decoder at a new B Origin' (in this example, B Origin 259);

- $M = 4 + 23$ — this implies 'suppress the first 4 digits of the incoming digit string and prefix the digits '23' to the string';

- $ES = 102$ — a termination condition implying 'digit decoding is complete for this call and the call is forwarded to the next processing stage within the exchange — the ES (end selection) value provides an index into that next stage';

- $RC = 19$ — a termination condition implying 'digit decoding is complete for this call — the call is forwarded to the next processing stage within the exchange — the RC (routeing case) value provides an index into that next stage'.

13.5.3   Traditional design

Adopting the traditional design approach, the team would have undertaken a data modelling exercise using the entity relationship diagram (ERD) tools of StP, Teamwork or other CASE tool. The results of such an exercise are presented in Fig. 13.8. Figure 13.8 might be simplified as shown in Fig. 13.9.

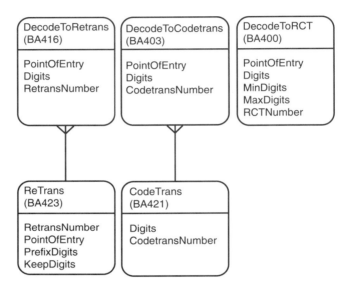

Fig. 13.8   Data modelling for System X digit decoder.

For AXE10, a similar data modelling exercise would have provided the model as shown in Fig. 13.10.

The result of the exercise would be two distinct data models, each of which are readily implemented with relational database tools. However, the data models are disjoint — they share no common components and the algorithms for tracing calls through the two models are also distinct. The scope for sharing common software design and implementation is limited and in practice the design and implementation of each exchange type within the model would be completed in isolation. Two data models and two algorithms represent (to an approximation) twice the development costs of an NRMS supporting only one exchange type.

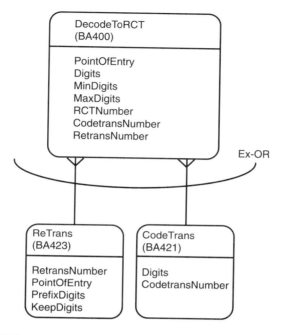

Fig. 13.9 Further data modelling for System X digit decoder.

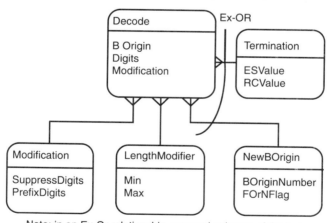

Note: in an Ex-Or relationship, one and only one of the two
crossed relationships will hold

Fig. 13.10 Data modelling for AXE10 digit decoder.

13.5.4   A generic digit decoder design

The task of the NRMS designers was to apply OO principles to the design of a generic digit decoder — one which would provide:

● significant class reuse between System X and AXE10;

● encapsulation of exchange type specific data structures and functionality;

● encapsulation of common, low-volatility, data structures and functionality for reuse between System X and AXE10 exchange types.

The digit decoder is modelled as a single class, a DAArea which comprises a set of decadic trees (class DATree). The model at this level provides a high degree of abstraction. Each tree corresponds to a 'point of entry' in System X or to a B Origin in AXE10 (see Fig. 13.11).

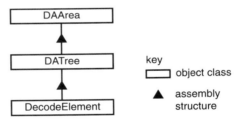

Fig. 13.11   Abstract structure of the digit decoder.

Each decadic tree has a basic building block — a DecodeElement class, which represents the diallable digits (0,1,...,9) at a given point in the digit string. Hence DecodeElements at the first level in the decadic tree deal with the first digit in the incoming digit string; those at the second level deal with the second digit, etc.

A digit position within a DecodeElement may point at a lower level DecodeElement, or at a DAResultSet class.

A DecodeElement itself may have a single DAResultSet associated. Figure 13.12 shows an example decode tree.

A DAResultSet holds a set of DAResources. Each DAResource encapsulates some processing (and related data) on the incoming digit string. Generically, three types of processing are supported.

● The first implies some manipulation of the digit string (digit insertion, digit deletion, etc). This class is DAModDigits. DAModDigits is further

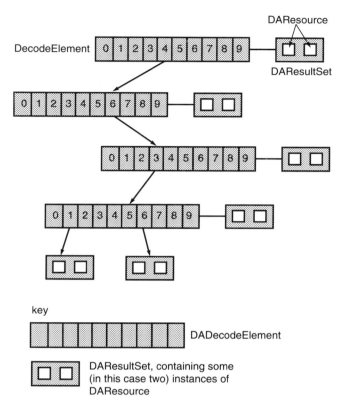

Fig. 13.12   Part of an example decode tree, showing the decode path for digit strings '4631' and '4636'.

specialized for each of the two exchange types to provide for exchange-specific string modification features.

- The second implies some onward routeing of the call within the exchange. The DADecodeComplete class will transfer a decoded call to the next processing stage in the exchange, either a routeing category (System X) or a routeing case or end selection (AXE10).

- The third implies returning the incoming digit string to the digit decoder to begin decoding again. This is handled by the DAReturnToDecoder class. This class handles the code translation and retranslation cases of System X whereby the call is returned with the same/different point of entry respectively. This class also handles the 'F = xxx' and 'N = xxx' retranslations found in an AXE10 exchange (see Fig. 13.13).

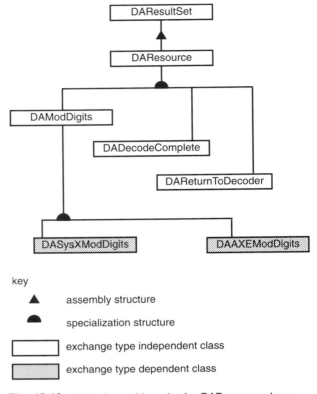

Fig. 13.13   Inheritance hierarchy for **DAResource** classes.

13.6   CONCLUSIONS

The lessons to be drawn at this stage in the NRMS product life cycle are fourfold.

13.6.1   Speed of initial product delivery is no faster

The use of OO techniques has not led to faster delivery to the field — the NRMS team has spent longer in analysis and design than it would have done using a traditional approach. The business of designing 'generic' classes is time-consuming. Had the team used traditional techniques the data modelling would have been carried out principally on the basis of normalizing the structures found in the BA tables and the AXE10 prints.

The pay-back for a greater investment in analysis and design is in reduced implementation time.

The total development time for NRMS by means of traditional or OO techniques is estimated to be similar.

13.6.2   Lifetime product costs are reduced

NRMS has been enhanced on three occasions since its original delivery in early 1992. Each enhancement has taken account of routeing data and functionality changes introduced by GPT for the System X exchange (for 4100, 4300 and 4500 builds).

In each case the cost of the enhancement can be compared with the cost of a similar enhancement to an existing BT-developed system that holds similar data and offers similar functionality.

A comparison of the existing deployed system with NRMS shows the cost of enhancements to NRMS are lower by a factor of between 2 and 4. NRMS replaced the existing system in mid-1993.

The abstract classes designed and implemented for NRMS provide encapsulations of the stable data and functionality of PSTN exchanges. The data and functionality which is more volatile has been encapsulated in a small number of classes. The impact of inevitable change has been localized and the cost contained to a relatively small part of the whole.

13.6.3   Accurate direction of resources

An interesting effect of the OO approach has been the fact that most effort has been directed at the design of the most stable classes within the system. The most stable classes are the generic classes.

The NRMS team comprises individuals who have studied, in-depth, either System X or AXE10 exchanges. In order to identify generic classes, all of the NRMS exchange experts collaborated in a single design exercise. Hence the design team for those classes was twice the size of the two teams that produced the exchange-type specific class designs.

13.6.4   Ease of communication

The OO paradigm — the concepts of classes with data and function, specialization/generalization structures, assembly structures, etc — is readily understood. Most importantly it can be readily understood by domain experts such as people with extensive expertise in exchanges. This contrasts with the less readily understood data-flow diagrams and structure charts of the past. The paradigm is easily understood by non-technical people, and therefore aids the vital communication between developer and customer.

REFERENCES

1.   Coad P and Yourdon E: 'Object oriented analysis', Yourdon Press (1992).

2.   Coad P and Yourdon E: 'Object oriented design', Yourdon Press (1992).

3.   ONTOS DB 2.2 Reference Manual, ONTOS Inc (1992).

4.   ONTOS DB 2.2 Developer's Guide, ONTOS Inc (1992).

5.   Software through Pictures, Reference Manual, Interactive Development Environments (IDE) (1992).

14

OBJECT ORIENTED CORRELATION

B M Osborn and C T Whitney

14.1   INTRODUCTION

This chapter describes how object oriented techniques have been applied to a real world problem, namely that of correlating transmission network alarms. An object oriented model of a significant proportion of BT's transmission network has been developed in order to perform this correlation. The alarm information is supplied by several systems monitoring the network in London. Correlation in this application means identifying the single probable cause of a group of alarms using information in the alarms themselves and in the model of the network. The correlation is performed in real time to provide identification of the root cause of the faults. To achieve this manually is an arduous and time-consuming task as the monitoring systems can generate a wealth of alarms as a result of a single network fault. This chapter described the development of a system to perform the correlation automatically.

Software techniques have been developed, termed 'out-of-model' and 'in-model' correlation. Both were developed using object oriented analysis, design and programming. The 'out-of-model' approach has the correlation process defined as a set of rules which use the network model as a passive source of data. The 'in-model' technique uses the objects in the network model itself and messages passed between them to perform the correlation. This second approach is based on work done with competence networks and has led to a system that is flexible and easy to extend and maintain.

The development of these software techniques has been driven by business requirements and has been realized using a prototyping approach. The validity and robustness of the software has been proved by its use in the field as part of the local fault management system (LFMS) feasibility trial.

This chapter is structured as follows:

- section 14.2 providing background to the transmission domain, the need for automatic correlation and the object oriented model which has been developed;

- section 14.3 explaining the two correlation techniques which have been used in the local fault management system;

- section 14.4 noting some early results from the feasibility field trial;

- section 14.5 concluding the paper by comparing the techniques discussed.

14.2   MODEL DEVELOPMENTS

14.2.1   The transmission network

At present the majority of the transmission network is plesiochronous, composed of a hierarchy of various types of equipment. These types include cables, multiplexers, regenerators, etc. Multiplexers enable a number of low-speed circuits to be carried over a single higher speed bearer. The multiplexed bearer is said to carry a number of tributary circuits. For example, a '2—8 mux' combines four 2 Mbit/s channels or tributaries to an 8 Mbit/s bearer. The 8 Mbit/s bearer can be thought of logically as four tributary circuits. A series of multiplexers, 2—8, 8—34, 34—140 and 140—565, allows construction of the multiplexed hierarchy shown in Fig. 14.1. There are also skip multiplexers that step up or down two or more transmission rate levels in one piece of equipment.

Multiplexing is used to reduce transmission costs by improving bearer media usage (a 565 Mbit/s bearer circuit can carry 7680 channels) and to decrease the number of bearer channels that need to be monitored and maintained, so improving the manageability of the network [1].

At present, different information about BT's network is held in separate databases in a variety of formats. For example, one database contains information about physical equipment in the field and another about customers. This situation has arisen as each database was developed with different requirements to support specific applications. When the network carried only a limited number of services and these did not change very

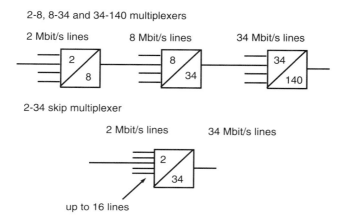

Fig. 14.1   The digital hierarchy.

frequently, this approach provided adequate support for the various network and service management needs. However, as both the range of services and of technologies increases and the rate of change of these services increases to meet the customer's expectations, there is a need for a generic information base. The network configuration model (NCM) is based on an interpretation of the managed objects defined by the Network Management Forum[1] (NMF) [2]. Managed objects are concerned with how to represent information across interfaces and to set standards for communication between different networks and services management systems. This communication between systems becomes increasingly important as BT focuses attention on network management through the use of computers. The co-operative networking architecture[2] (CNA) uses international standards where available to specify the basis for open systems procurement and development. The NCM takes data from several sources and presents it in a CNA conformant format for applications such as the local fault management system (LFMS).

The alarm information of fault reports comes from three major systems. One monitors private circuits (30% of the 1 million private circuits are in

---

[1] The Network Management Forum is an industry consortium of leading international computing and network equipment suppliers, service providers and users. The document referred to here sets out definitions for interoperable communications network management products that support alarm management. The architecture is based on ISO standards and CCITT recommendations.

[2] The co-operative networking architecture group administer a library of NMF managed object classes and extensions registered for use within BT [3].

London [1], so this provides a rich source of alarm data). The second source monitors circuits at all transmission rates and gives a wide range of different alarms. The third system includes alarms against PSTN 2 Mbit/s circuits.

Network equipment can develop faults. These faults include loss of synchronization, power failure and degraded signal, for example. The pieces of equipment will send out alarm signals which are converted to alarm reports by the monitoring systems mentioned above. Other pieces of equipment, dependent on an item with a fault, will also register alarm messages as they can no longer function. Further pieces dependent on these may also send alarms and so on. Consequently, the failure of one higher order piece of equipment can result in a deluge of alarms. It can be difficult to identify quickly and accurately the cause of all the alarms. A further complication is that not all equipment is monitored, so a complete set of alarms is unlikely. Also the use of the different technologies means that not all alarms pertaining to a particular fault will be received at the same time. These difficulties provided the impetus for the development of the LFMS.

14.2.2   The model

The transmission domain was an ideal candidate for an object oriented representation. There were clearly-defined 'objects' in the domain, both logical and physical. The object oriented approach allowed these objects to be represented as naturally as possible. It is a method which 'appeals to the workings of human cognition' [4].

Functions and data are treated as two indivisible aspects of objects. The design can be expository, i.e. setting forth a view of the world, rather than synthetic, as components of a program are often electronic realizations of entities (physical or recognized abstractions) in the real world [5].

According to Halbert and O'Brien: 'In control-based programming, you are forced to modularize programs based on procedural criteria. This often leads to program structures that are radically different from the structures in the application domain. In object oriented programming, types (classes of objects) provide a natural basis for modularization in a program because they are commonly used to model entities in the application domain. Because types model application entities, the structures in object oriented programs can more closely resemble the structures in the application domain' [6].

The analysis and design followed Booch's methodology [4]. Firstly, the classes and objects in the domain were identified, followed by the semantics of these objects and classes. Then the relationships among the classes and objects were identified. These classes and objects were implemented. The process was repeated until no new abstractions could be found. The objects in the domain were the logical ones of circuits, function and service, and the physical ones of equipment, facility and location.

The LFMS is part of a larger area of work within BT (see Smith, Butler and Azarmi [7]). It is shown diagrammatically in Fig. 14.2. One part is the development of a model, the network configuration model (mentioned in section 14.2.1), of the plesiochronous digital hierarchy (PDH) transmission network.

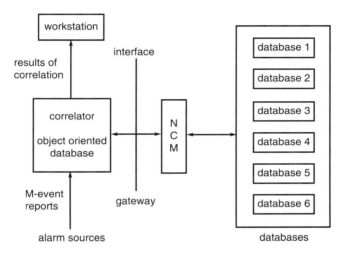

Fig. 14.2 Overview of LFMS and related projects.

The initial classes for the LFMS were those chosen by the NCM project team as the most appropriate for the network model from the NMF managed objects. These are shown with their relationships in Fig. 14.3.

The transmission network is modelled in the following way. A circuit has one or more componentNames giving its routeing over other circuits. It terminates on two function instances. The function instances are provided by one or more instances of equipment. The circuit is carried on one or more instances of facility which terminate on instances of equipment. The instances of equipment reside in a location. The circuit, if it is a private circuit, provides a service.

Some of the NMF classes have attributes in common. All the managed object classes chosen have the attributes userLabels and typeText. The function, circuit and facility classes share the attributes administrativeState and operationalState. The circuit and facility classes share the transmission Direction attribute. So the first step was to design an inheritance hierarchy to factor out these common attributes. The managed object library is not

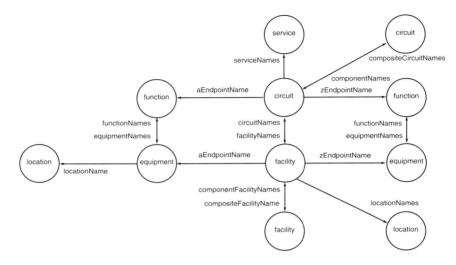

Fig. 14.3   Relationships between mnaged object classes.

intended to restrict the way in which data is structured within the applications using it. It is defined only for the purposes of communication. Thus there was the freedom to design superclasses to eliminate duplication of attrributes and their associated services.

Abstract superclasses, from which the common attributes could be inherited, were created. Each abstract class (see Chapter 1 for a definition of abstract classes, i.e. classes that are designed not to have immediate instances) has methods to return, change and display the value of its attribute or attributes (see Fig. 14.4):

- userLabels class has attribute userLabels;

- typeText class has attribute typeText;

- states class has attributes administrativeState, operationalState;

- cirfac class has attribute transmissionDirection.

Additional relationships between the classes were added in order to maintain referential integrity in the persistent store. The entity relationship diagram shown in Fig. 14.5 shows all the relationships between instances in the model. (A comparison of Fig. 14.5 and Fig. 14.3 will show the extra relationships that have been added). Where the NCM model has pointers in both directions, these have been implemented but where that model lacks a reverse direction pointer, for example, between location and equipment, one has been implemented, in order to maintain the referential integrity

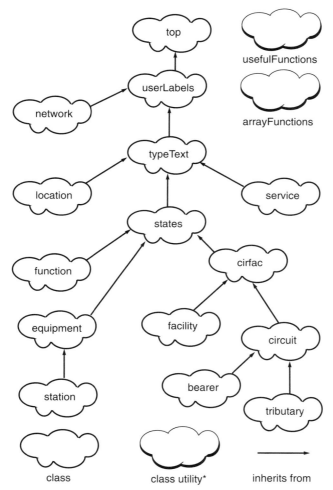

Fig. 14.4   Inheritance hierarchy.

of the model. For example, suppose instance *A* of the equipment class has a pointer or reference to instance *B* of the location class. If instance *B* is deleted, then instance *A* will be left with an invalid or dangling pointer. The relationship locationNameBack allows the equipment instance to be found by the location instance before it is deleted and the pointer can be set to null.

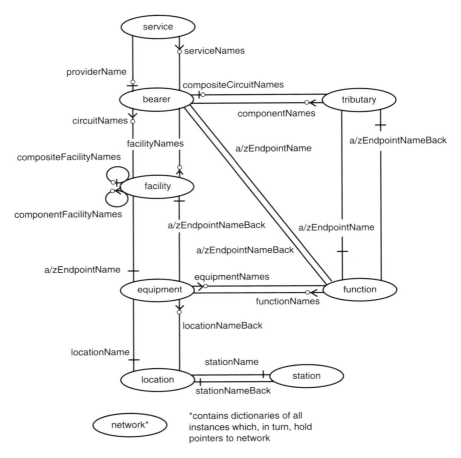

The a/zEndpointName and a/zEndpointNameBack relationships are implemented between circuit and function rather than bearer and function and tributary and function.

Fig. 14.5   Entity relationship diagram.

In the first database used for implementation of the model, the code to maintain integrity has to be written by the developer. In the second database in which the model was implemented, as long as an association is declared as bidirectional, integrity is automatically maintained.

Further subclassing was introduced to specialize circuits and equipment. Bearer and tributary are subclasses of circuit. These two classes are used to enable the correlator to navigate through the hierarchy of circuits, as follows. A circuit can be thought of as a bearer circuit sharing its bandwidth between lower order circuits. For example, an 8 Mbit/s circuit is a bearer for up to four 2 Mbit/s circuits. Each 2 Mbit/s circuit will have in its componentNames attribute the appropriate tributary of the 8 Mbit/s circuit. Each tributary of the 8 Mbit/s circuit will have the name of the 2 Mbit/s circuit that it is carrying in its compositeCircuitNames attribute. In turn the 8 Mbit/s circuits will have the tributary of a 34 Mbit/s circuit in its componentNames. (NCM uses the managed object class 'circuit' to model bearers and tributaries.) Owing to restrictions in the definitions of the managed object classes, it was decided to use the 'equipment' class to model stations or exchanges in the NCM model; therefore the equipment class has been subclassed to provide a 'station' instance in the LFMS model.

An often-quoted advantage of object oriented programming, that the implementation of methods operating on data is hidden from using systems, was realized when the model was re-implemented in a different object oriented database. The database specific code within functions had to be rewritten and minor changes, like the syntax of starting a database transaction, had to be made but the large majority of the code remained unaltered. This not only saved considerable time and effort for the LFMS developers but also demonstrated that the system is not tied to one particular object oriented database vendor.

14.2.3 The local fault management system project

The LFMS aimed to demonstrate that the correlation of a variety of alarm sources using an object oriented model of the transmission network was both feasible and beneficial. It was previously thought that real-time correlation based on a network model was not feasible. Other aims included the reduction of alarm information presented to network managers and the adding of value to the information, the support of more rapid and accurate identification of the cause of alarm reports to facilitate speedy repair, and the provision of reliable information to those dealing with customer queries.

Initial requirements capture was undertaken by visiting several network operations units. This enabled the developers to interview network managers and to see their methods and the problems they experienced first hand. The prototyping development approach allowed requirements capture to continue. (For a discussion of the use of prototyping in the development of object oriented systems, see Mullin [5].) At the end of each prototype iteration the end users were invited to comment formally on the system. As the developers', and in some cases the users', knowledge of the domain increased, changes were made to the correlation process and to the way the results were presented. Many enhancements were made as a result of the users' suggestions and requests.

Many argue that a high-level object oriented language, for example Smalltalk, should be chosen to develop a prototype. However C++ was chosen as the development language for LFMS. This was partly because of the intention to use an object oriented database which is a persistent store of C++ objects. Also a high-level language lends itself well to prototyping interfaces, whereas with this project the aim was to understand and demonstrate the required functionality. The functional requirements included a system that could run continuously. It needed to be able to interface easily to other systems, in order to receive incoming alarms and to transmit correlation reports. The system was also required to operate in real time. In section 14.4 some of the benefits of using C++ are reported.

14.3   CORRELATION TECHNIQUES

The objective of correlation is to reduce the amount of information passed on to the network managers and to add value to this information. After correlation, the raw fault reports should have been processed into fewer, richer correlation reports, allowing maintenance engineers to identify necessary repairs and prioritize them.

14.3.1   Introduction

Correlation is a means of analysing data and finding commonalities contained within the data. As such, it is a powerful mechanism for investigation and diagnosis which has been applied to many domains, ranging in diversity from network fault isolation to financial trend analysis. Correlation involves identifying attributes that are common to a set of data and subsequently using these attributes to identify further attributes in the data. This process continues until either no more commonalities can be found or a successful diagnosis has been achieved.

There are many types of correlation largely defined in terms of the commonality in the data that is being identified. The most common types are [8]:

● spatial correlation, where physical location of data or physical locations described in the attributes of the data are correlated — an example of this type of correlation would be 'all these alarms were generated by equipment on the same shelf in this exchange';

- temporal correlation (see Chapter 8), where the temporal appearance of the data or temporal information described in the attributes of data are correlated — an example of this type of correlation would be 'these alarms all happened within the same 5 minute period';

- connective correlation, where common associations between data are correlated — an example of this type of correlation would be 'the last three alarms all orignated from circuits that are carried by this higher order circuit'.

Correlation is, by its very nature, data-intensive and any means of structuring the data or using the data in a formal framework will greatly ease the process. The object oriented paradigm provides such a framework:

- data can be structured by attributes via class definitions;

- data can be structured hierarchically via inheritance;

- dependencies between data can be specified via association;

- mechanisms to search data for commonalities can be specified via methods.

By using the object oriented paradigm, data and the process of correlation of the data can be much more tightly bound.

Important to the correlation process is how long an alarm of a correlation should be deemed valid and appropriate for use for correlation. If alarms and correlations last forever, then eventually every circuit in the network would be on alarm and the whole LFMS model would be correlated.

The correlation process must realize that alarms and correlations only have a certain lifetime in which they are useful and after which they should be ignored or removed. As shown in Fig. 14.6, a correlation window has to be defined.

A correlation window is the amount of time that elapses before sets of alarms are correlated. The lifetime of an alarm must be approximately the same length of the correlation window so that alarms do not appear and disappear without being correlated. Alternatively the lifetime of an alarm should not be so great that it persists in many correlation windows, possibly giving false results. User feedback was very important in finding the optimum length for this window.

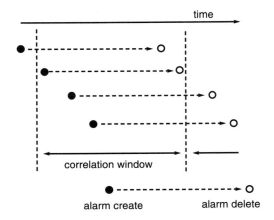

Fig. 14.6   Correlation window.

14.3.2   Use of the model for correlation

The LFMS project is initially concerned with carrying out connective correlation on the PDH network and, as this chapter describes, it takes full advantage of the benefits provided by object orientation in order to achieve this. Connective correlation within the LFMS correlator is performed by exploiting the Carried-By and Carries associations between circuit objects contained within the model.

A Carried-By association exists between two circuit objects A and B if:

- A is a lower order circuit;
- B is a higher order circuit;
- A is carried by B, i.e. A is a circuit routed over a tributary of B.

The Carried-By association is a one-to-many relationship, i.e. a lower order circuit can be carried by more than one higher order circuit.

Figure 14.7 shows the Carried-By association contained within the model. The association identifies a set of higher order circuit objects that realize a lower order circuit object. It is this association that provides the structure to the LFMS model as it defines how circuits are physically related in the actual PDH network. For example, if two 2 Mbit/s circuits are routed over an 8 Mbit/s circuit in the actual PDH network then there will be Carried-By associations between the 2 Mbit/s and 8 Mbit/s circuit objects representing these circuits in the LFMS model. In the same way, if two 8 Mbit/s circuits

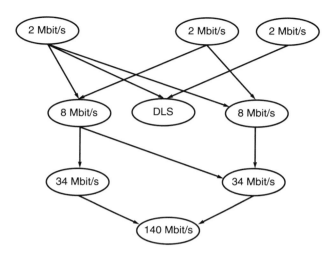

Fig. 14.7 The Carried-By association.

are routed over a 34 Mbit/s circuit in the actual PDH network, then there will be associations between the correct 8 Mbit/s and 34 Mbit/s circuit objects. It can be seen that it is these Carried-By associations that provide the LFMS model with its strict hierarchical structure. Every circuit object is related to other circuit objects by means of the Carried-By association.

The LFMS correlator is designed to identify and report about higher order circuit objects that are associated, via the Carried-By association, with lower order circuit objects on alarm. The rationale behind this is that if two circuits are on alarm and these two circuits are realized by higher order circuits that are common to those circuits on alarm, then it may be that these common circuits are at fault and should therefore be reported or maybe become candidates for further correlation.

An inverse association to the Carried-By association exists in the model. Known as the Carries association, it exists between two circuit objects *A* and *B* if:

- *A* is a higher order circuit;
- *B* is a lower order circuit;
- *A* carries *B*, i.e. *A* has *B* routed over one of its tributaries.

The Carries association is a one-to-many relationship, i.e. a higher order circuit can carry many lower order circuits. The association identifies a set of lower order circuits that are carried over a higher order circuit. For example, if an 8 Mbit/s circuit carries two 2 Mbit/s circuits in the actual PDH network, then there will be Carries associations between the circuit objects that represent these circuits in the LFMS model. In the same way, if a 140 Mbit/s circuit carries two 34 Mbit/s circuits in the PDH network, there will be Carries associations between the respective objects in the PDH model. (As the Carries association is the inverse of the Carried-By association, it is not illustrated.)

The Carries association is also used in the correlation process but in a manner that is not obvious. However this will become clearer later in this section. Additionally the Carries association is used after a correlation has been produced to identify all the potentially affected circuits.

The Carried-By and Carries associations are actually implemented (as described in section 14.2.2) as relationships between bearer and tributary instances. The class methods and the methods in the class utilities hide the actual implementation. An external application can request all the circuits routed over a particular higher order circuit. The methods will navigate the appropriate bearer and tributary circuits invisibly.

14.3.3   Out-of-model correlation

The initial approach to performing correlation with the model was to carry out what can be termed 'out-of-model' correlation. This is so called because the correlation process is separated from the model. Information is retrieved by querying the model in much the same way aas a conventional database as shown in Fig. 14.8.

The correlation process itself is defined as a set of rules which operate in a forward-chaining manner upon information retrieved from the model.

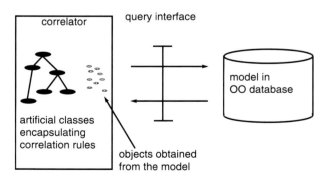

Fig. 14.8   Out-of-model correlation process.

The rules themselves are implemented as methods in a set of artificial classes. Rules are triggered by the arrival of alarms and the production of correlations to produce new correlations and correlation reports. The firing of rules continues until no more correlations can be found or all possible correlation reports have been produced or until new alarms arrive for correlation. A high-level description of a correlation rule would be:

```
IF          alarm1 is a new alarm
THEN        retrieve circuit object for alarm1 from the model
```

This describes the situation where a new alarm has arrived and information about the circuit that is on alarm is retrieved from the database. Another example would be:

```
IF          circuit1 is a circuit object retrieved from the model
    AND     circuit2 is a circuit object retrieved from the model
    AND     parents1 is the routeing information for circuit1
    AND     parents2 is the routeing information for circuit2
    AND     intersection of parents1 and parents2 is not empty
THEN
            circuit3 is the highest common circuit in the intersection
    AND     retrieve circuit object for circuit3 from the model
```

This describes the situation where two circuit objects have been retrieved from the model and they share a common parent. The common parent is a candidate for further correlation and so its information is retrieved from the model.

Initial rules filter out alarms from circuits that should not be passed forward for correlation. An example would be an alarm from a circuit identified in the database as being in course of provision.

The rules themselves are implemented in C++ as methods of artificial classes for use by the correlation process only. The classes are artificial in the sense that they do not describe objects in the real world. They exist only to assist in the correlation of objects from the real world, i.e. alarms and circuits.

It was useful at first to keep the model and the correlation process separate in order to check the validity of the correlation process. However, it soon became obvious that the 'out-of-model' correlation approach described above was inappropriate in the long term. Separating the model from the correlation process prevented any use being made of benefits that could be derived from a model of the domain. The model was reduced to a passive role providing information about the domain when requested and playing no further role in the correlation process. It can be argued that the use of an object oriented database for this purpose is inappropriate and that a conventional database would have been more suited to the task.

Implementing rules as methods of a set of artificial classes is also inappropriate. One of the advantages of adopting a rule-based approach to problem solving is that the problem-solving knowledge can easily be expressed, visualized and modified when encapsulated in an IF/THEN rule format. It is from this expressibility that most expert system shells derive their power. However, to describe the problem-solving knowledge for correlation as a set of rules and then to implement these rules as a set of methods negates any of the benefits that can be derived from a rule-based approach. C++ is not a language for visualizing problem-solving knowledge and makes modifying this knowledge difficult and prone to error.

It was for the above reasons that an alternative approach to correlation was adopted — the so called 'in-model' correlation approach described in the next section.

14.3.4   In-model correlation

A major influence upon the design of the 'in-model' correlator and one that is supported by the object-oriented paradigm is that of distributed intelligence. More specifically, the design has many ideas borrowed heavily from the work of Maes [9] and her theory of competence networks.

There has been much activity recently in the field of artificial intelligence (AI) with the emphasis on building systems composed of many interacting, simple behaviour-producing elements which, when combined, provide the system with its overall intelligence [10]. This is opposed to the more traditional approach of classical AI of building systems with clearly defined information processing modules, the combination of which provides overall intelligence. Supporters claim that systems built around the notion of distributed intelligence overcome many of the problems, such as inflexibility, brittleness and lack of real-time performance, that are prevalent in systems built in classical AI frameworks.

A good example of how to build systems with distributed intelligence has been provided by Maes [9]. The algorithm was primarily designed to provide a mechanism for action selection in autonomous agents. However, many of her ideas are applicable to other applications such as connective correlation. This is true both for LFMS correlation and for other domains that utilize an object oriented model. The theory is based around a collection of simple behaviour-producing components referred to as competence modules. These are so called because each module has a simple behaviour that will be exhibited when the module is activated.

Firstly, a parallel is drawn between competence modules and circuit objects in the LFMS model. Competence modules are designed to exhibit very simple correlation behaviour when activated by associating appropriate methods to the objects. Competence modules are expressed as the tuple (c, a, d, i):

- c — a list of pre-conditions that must be fulfilled before the module becomes active;

- a, d — the expected effects of the competence module's action in terms of an add and delete list of fulfilled conditions;

- i — the level of activation for a module.

The above can be interpreted as 'when my energy reaches my activation level and all my pre-conditions have been fulfilled, then I will carry out an action that will result in the conditions of my add/delete lists being fulfilled'. Energy is introduced into a module along the pre-conditions list as described later. For the purposes of correlation, circuit objects can also be expressed as a tuple (c, a, i):

- c — a list of incoming associations relevant to correlation, i.e Carries associations;

- a — a list of outgoing associations relevant to correlation, i.e. Carried-By associations;

- i — activation level for the object.

In Maes' competence networks, the modules form a network by virtue of three types of link:

- successor links, between the add and pre-condition lists of modules;

- predecessor links, the inverse of successor links;

- conflictor links, between the delete pre-condition lists of modules.

These links define how energy flows around the competence network to obtain the desired behaviour. Similarly circuit objects are linked into a correlation network by means of the incoming and outgoing associations contained in the typle. This correlation network is implicit in the LFMS model as the incoming/outgoing associations correspond to the Carries/Carried-By associations already present in the circuit objects. It is merely a matter of designing the correct methods for the circuit objects to utilize the associations so as to obtain the desired behaviour.

Correlation proceeds by injecting energy into the network and using the associations to propagate energy forwards and backwards through the network. Energy is expressed as a numerical value and is given to circuit objects that are currently on alarm as it is the activation of the circuit alarms that provides the impetus for correlation. Propagation of energy from the circuit objects on alarm is carried out along both the Carried-By and Carries associations:

- Carried-By — energy is passed along this association (forwards in the model) in an attempt to correlate higher order circuits with lower order circuits on alarm;

- Carries — energy is passed backwards along this association (backwards in the model) in an attempt to activate lower order circuits that have not yet been activated but that should have been, based on the fact that other lower order circuits have been activated and have passed energy forwards.

As described previously, alarms and correlations are only useful for correlation for a predetermined length of time, an alarm's correlation window, after which time they can be discarded. In order to reflect this in the LFMS correlator both the energy injected into the network by alarms and the energy that accumulates in objects are subject to decay governed by a decay function which ensures that the amount of energy decreases to zero with respect to time as shown in Fig. 14.9.

At the start of an alarm's correlation window it will introduce the maximum level of energy possible into the network. This represents the fact that the alarm is at its most relevant for correlation at this point. As time passes, the relevancy of an alarm with respect to correlation decreases. This is represented by the decaying of an alarm's energy in the network. At the end of an alarm's correlation window it is no longer useful. This is represented by the fact that at this time the amount of energy in the network for this alarm should be zero.

An 8 Mbit/s correlation based on two 2 Mbit/s alarms is more relevant than a 34 Mbit/s correlation based on the same two alarms. Similarly a 34 Mbit/s correlation is more relevant than a 140 Mbit/s correlation based on the alarms. Generally the closer the correlation is to the source of the alarms the more relevant the correlation. As a correlation is produced when the

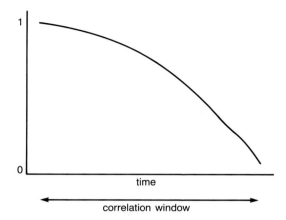

Fig. 14.9   Decay function.

activation level for an object crosses a threshold, relevancy can be ensured by only passing a percentage of the activation energy forwards along the Carried-By associations. This ensures that higher order correlations are only produced when many of the lower order circuits that it carries are on alarm or have correlated. The amount of energy that a circuit object passes forwards will always be proportional to the number of circuits it carries.

In her algorithm, Maes [9] only activates a module when:

- all its pre-conditions have been fulfilled;
- its activation level crosses a threshold.

In the application described here this would amount to an object having to wait for:

- all circuits below a higher order circuit to go on alarm or all circuits below to have activated;
- its activation energy to cross a threshold.

Not all circuits in the PDH network are monitored for alarms so it cannot always be guaranteed that the first condition will be met for all circuits. To avoid this incompleteness problem the first requirement for the activation of an object has been dropped. An object will activate purely on its activation energy crossing a threshold. However it can still be guaranteed that a circuit only activates when appropriate by setting the thresholds for activation of an object to suitable values.

An example will illustrate how an 8 Mbit/s correlation can be produced from two 2 Mbit/s alarms. It will then be shown how a 34 Mbit/s correlation can be produced by a further 2 Mbit/s alarm. A fragment of the model is shown in Fig. 14.10.

The arrival of the first alarm results in the 2 Mbit/s circuit object passing energy forwards to all 8 Mbit/s circuit objects that carry it. The 8 Mbit/s circuit objects in turn pass energy forwards to all 34 Mbit/s circuit objects that carry them, and so on. When the second alarm arrives at a 2 Mbit/s circuit object again energy is passed forwards to all 8 Mbit/s circuit objects that support it. However, this time the energy in one of the 8 Mbit/s circuit objects crosses its threshold and results in its activation producing a correlation report. By now the 34 Mbit/s circuit object has received two bursts of energy but still has not passed its threshold because it has not got enough energy. It passes some of its energy back to the 8 Mbit/s circuit object that it carried to prompt them to activate and pass it more energy. The arrival of the third alarm at a 2 Mbit/s circuit object results in energy being passed forwards to first the 8 Mbit/s circuit object and then the 34 Mbit/s circuit object. This time the 34 Mbit/s circuit object has enough energy to activate and results in a correlation report being produced.

Once a correlation report has been produced, an affected feature report is also produced. This report lists all the circuits that are routed over the

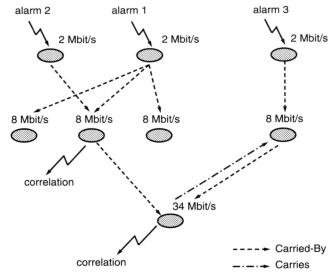

Fig. 14.10   Correlation in the model.

higher order circuit identified in the correlation report, i.e. all the circuits affected by this fault. The circuits are separated into groups by importance. Initially only those circuits belonging to private customers were recorded but as a result of user feedback, all circuits were included, separated into customer, signalling and general PSTN circuits. It is generated by using the Carriers association. The report enables network operators to prioritize repair work and to respond to customer queries about problems with their circuits.

14.4   FIELD TRIAL RESULTS

The field trial is demonstrating a large reduction factor between alarms received and correlations generated. As a result, people managing the network are presented with a much smaller volume of information but of significantly greater value. Major network failures were identified well in advance of current practices. Also there were correlations based on transient faults generated several days in advance of failures of higher order systems. This would enable engineers to become proactive, preventing failures as well as repairing faults when they occur. The reduction in the amount of data presented to end users allowed the possibility of analysis of data not previously attempted. For example, improvements were made to the operational procedure for ceasing private circuits when customers no longer require them. The success of the system is illustrated by the fact that users requested a significant extension to the field trial period and results are beginning to be used operationally.

The advantages of using C++ became evident during the initial software development and subsequently as modifications and enhancements were made. The user interface to the LFMS was built using a proprietary BT package using C++. The ease with which attributes and behaviours of existing classes can be inherited to developer-designed classes was exploited to make the frequency modifications to the interface requested by the end users.

C++ was again beneficial in the unpacking of the raw alarm data. A class to interpret alarms into a format useful for correlation was subclassed to deal with the many different formats of incoming data. This resulted in code that was both easy to understand and straightforward to extend for new formats. C++ classes were also designed and used for the communications between processes in LFMS and to gather metrics. Again simple subclassing made the same software easily reusable in differing situations, and indeed some of the software was reused in another project being undertaken in the same group.

The LFMS coped with the demands of a real-time system, dealing with several alarms per second at peak rates and activating many hundreds of objects in the model. C++ was important here in allowing this performance. Also careful memory management was significant for a system running continuously.

The affected feature report, listing private circuits potentially affected by network faults identified by the correlation process, proved to be particularly useful and was extended at the users' request to cover signalling and other PSTN circuits.

14.5   CONCLUSIONS

Significant results have been achieved in a short space of time by the application of object oriented analysis, design and programming techniques. A comprehensive model of a substantial part of BT's transmission network has been produced using internationally agreed standards. It will readily allow further development and change. It was implemented in a natural way in two commercially available object oriented databases. The reimplementation of the model in the second object oriented database required only very limited and readily identifiable software modifications, highlighting the value of encapsulation.

The 'in-model' approach developed more recently follows much more closely the object oriented philosophy. Objects are given attributes, behaviour and state [4] (see Chapter 1 for an explanation of these terms). The attributes are defined in class definitions and represent facets from the domain. Behaviour is represented as a set of methods and defines how the objects should behave when carrying out the correlation within the domain. The concept of state is inherently realized using this approach because the values of an object's attributes vary over its lifetime whilst it operates within the domain. The 'in-model' approach to correlation has in-built flexibility for change. This will allow more data to be added to the model and additional correlation strategies to be included without the need for fundamental software changes. The design elegance leads to a degree of simplicity which will aid further software maintenance — an important aspect to be borne in mind.

The use of a Maes-style algorithm to define the correlation process has worked very well in this domain and it is anticipated that it would work well in other domains where similar types of correlation have to be performed. The algorithm fits well with the object oriented philosophy. Competence modules can be represented as objects, pre- and post-conditions can be represented as associations between objects, and the flow of energy can be represented as the passing of messages along the associations between objects. The flow of activation energy throughout the model fits well with what actually happens in the domain, that is, for example, a 2 Mbit/s fault may cause an 8 Mbit/s alarm which in turn may cause a 34 Mbit/s alarm. Another important feature of the Maes approach is that it provides an excellent opportunity to scale the domain via the use of parallel processing. Within the algorithm each object adopts a life of its own, operating independently. There is no reason why sets of objects cannot reside and operate in separate processors provided that there are appropriate communication channels to allow objects to communicate with each other.

User feedback suggested many valuable ideas for extensions and changes to make the system a better fit to their operational needs. The prototyping approach to software development has met this challenge by allowing enhancements to be made very rapidly. The object oriented approach is now showing its maturity and it fits well with an incremental software development and delivery strategy which allows end users to provide early feedback.

REFERENCES

1. IBTE: 'Telecommunications engineering — a structured information programme', (supplement supplied in instalments with each edition of the journal 'British Telecommunications Engineering') (1991).

2. OSI/Network Management Forum: 'Library of Managed Object Classes, Name Bindings and Attributes', Forum 006 Issue 1.1 (1990).

3. CNA Secretariat, Bibb Way, Ipswich, England, IP1 2EQ.

4. Booch G: 'Object oriented design with applications', Benjamin Cummings Publishing Company (1991).

5. Mullin M: 'Rapid prototyping for object oriented systems', Addison-Wesley (1990).

6. Halbert D C and O'Brien P D: 'Using types and inheritance in object oriented programming', IEEE Software (September 1987).

7. Smith R, Butler J and Azarmi N: 'A research and development programme for equipment and network diagnosis', BT Technol J, $\underline{11}$, No 1, pp 79-85 (January 1993).

8. Stein W and Cavanagh P: 'Correlation definitions', Internal BT report (1991).

9. Maes P: 'Situated agents can have goals', in Maes P (Ed): 'Designing Autonomous Agents: Theory and Practice from Biology to Engineering and Back', MIT Press (1990).

10. Brooks R: 'Elephants don't play chess', in Maes P (Ed): 'Designing Autonomous Agents: Theory and Practice from Biology to Engineering and Back', MIT Press (1990).

Index